Foreword

Some light end materials, such as LNG or LPG, have a bad safety reputation because of a few famous incidents such as Cleveland, Feyzin, Los Alfaques or Mexico City. However, when handled professionally by well-trained personnel in correctly designed installations, they present no more risk than any other energy sources. This booklet is intended for those operators, engineers and technicians working on process plant in order to raise awareness and promote safe designs and practices to avoid the occurrence of such incidents.

I strongly recommend you take the time to read this book carefully. The usefulness of this booklet is not limited to operating people; there are many useful applications for the maintenance, design and construction of facilities.

Please feel free to share your experience with others since this is one of the most effective means of communicating lessons learned and avoiding safety incidents in the future.

Greg Coleman, Group Vice President, HSSE

Acknowledgements

The co-operation of the following in providing data and illustrations for this edition is gratefully acknowledged:

- BP Refining Process Safety Network
- Albert Steiner and Pippa Ross, BP LPG Europe
- Neil Macnaughton, BP Grangemouth Refinery
- John Frame, Resource Protection International
- Keith Wilson, IChemE Loss Prevention Panel member

safety
sharingtheexperience
improving the way lessons are learned
through people, process and technology

BP Process Safety Series

Safe Handling of Light Ends

A collection of booklets describing hazards and how to manage them

This booklet is intended as a safety supplement to operator training courses, operating manuals, and operating procedures. It is provided to help the reader better understand the 'why' of safe operating practices and procedures in our plants. Important engineering design features are included. However, technical advances and other changes made after its publication, while generally not affecting principles, could affect some suggestions made herein. The reader is encouraged to examine such advances and changes when selecting and implementing practices and procedures at his/her facility.

While the information in this booklet is intended to increase the store-house of knowledge in safe operations, it is important for the reader to recognize that this material is generic in nature, that it is not unit specific, and, accordingly, that its contents may not be subject to literal application. Instead, as noted above, it is supplemental information for use in already established training programmes; and it should not be treated as a substitute for otherwise applicable operator training courses, operating manuals or operating procedures. The advice in this booklet is a matter of opinion only and should not be construed as a representation or statement of any kind as to the effect of following such advice and no responsibility for the use of it can be assumed by BP.

This disclaimer shall have effect only to the extent permitted by any applicable law.

Queries and suggestions regarding the technical content of this booklet should be addressed to Frédéric Gil, BP, Chertsey Road, Sunbury on Thames, TW16 7LN, UK. E-mail: gilf@bp.com

All rights reserved. No part of this publication may be reproduced, stored in a retrieval system, or transmitted, in any form or by any means, electronic, mechanical, photocopying, recording or otherwise, without the prior permission of the publisher.

Published by
Institution of Chemical Engineers (IChemE)
Davis Building
165–189 Railway Terrace
Rugby, CV21 3HQ, UK

IChemE is a Registered Charity in England and Wales
Offices in Rugby (UK), London (UK), Melbourne (Australia) and Kuala Lumpur (Malaysia)

© 2007 BP International Limited

ISBN-13: 978 0 85295 516 1

First edition 1961; Second edition 1964; Third edition 1984; Fourth edition 2005; Fifth edition 2007

Typeset by Techset Composition Limited, Salisbury, UK
Printed by Henry Ling, Dorchester, UK

Contents

1	**Introduction**	1
2	**Light ends defined**	4
3	**ABC's of light ends**	6
3.1	As a gas	6
3.2	As a liquid	8
4	**Light ends are everywhere**	17
5	**Handling light ends on process units**	18
5.1	During start-up	19
5.2	During shutdown	19
5.3	During processing	21
6	**Storage of light ends**	26
6.1	Activating storage vessels	28
6.2	De-activating storage vessels	30
6.3	Operations during active service	34
7	**Handling LPG at truck transport and railroad car terminals**	36
7.1	Operations common to loading and unloading tank cars and tank trucks	37
7.2	Operations concerning truck transports	43
7.3	Operations concerning railroad tank cars	44
8	**Design considerations**	46
8.1	Materials	46
8.2	Equipment	49
8.3	Fire protection	62
8.4	Design considerations for loading facilities	66
9	**Handling of LPG and light ends emergencies**	67
9.1	Behaviour of light ends spills and fires	68
9.2	LPG incident response strategies	69
10	**Some points to remember**	86

Bibliography **90**
**Appendix 1: Example of operator fire safety
 checklist for LPG storage** **92**
Test yourself! **95**
Acronyms and abbreviations **96**

NOTE: All units in this booklet are in US and metric systems.

1

Introduction

Skill and knowledge are required to handle all hydrocarbons whether at the well head, on a process unit or as finished products. However, one group of these hydrocarbons, known as *light ends*, has proven to be particularly hazardous. Experience has taught that only well-trained personnel, using properly designed equipment, can handle them safely.

Light ends are more difficult to contain in equipment than heavier hydrocarbons and are more hazardous if allowed to escape. The low viscosity of light ends (Figure 1), compared to heavier hydrocarbons, greatly aggravates the problem of containing them.

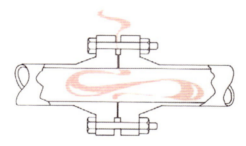

Figure 1 Light ends are more difficult to contain than heavy hydrocarbons.

Most of these hydrocarbons normally are gases at atmospheric temperature and pressure. To handle them as liquids, they must be confined under pressure or be held at low temperatures, or both. If the liquid leaks from a container, it will quickly vaporize, mix with air (oxygen) and probably develop a flammable mixture. The hazard here is increased because most light ends are heavier than air and will spread along the ground (Figure 2) where there are many possible sources of ignition (refer to Chapter 3 for more details).

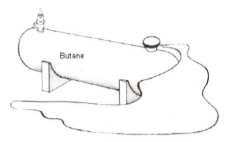

1 cubic foot (7.5 gallons/28.3 litres) of liquid = 225 cubic feet (6.3 m³) of gas

Figure 2 *Most light-ends vapours are heavier than air.*

Propane's ratio of gas volume to liquid volume at standard temperature and pressure is approximately 270:1, and for butane the figure is 225:1.

Serious fires and explosions have occurred because light ends have not been contained in refinery units (Figure 3), in storage tanks (Figure 4), at loading or unloading facilities (Figure 5a), during transport (Figure 5b) or at end-user locations (Figure 5c).

ACCIDENT

Figure 3 This fire at a refinery catalytic cracking unit started with a propane leak. The white circle on bottom left indicates the position of a fire truck where seven firefighters were killed.

Figure 4 Fire at refinery storage area. Eighteen firefighters were killed during this incident, when the first sphere BLEVEd.

Figure 5a This loading-rack fire resulted when the truck driver forgot to disconnect his fill hose before driving away.

Figure 5b This bulldozer hit a gas pipeline. The driver was killed by the resultant fireball.

continued

SAFE HANDLING OF LIGHT ENDS

Figure 5c *This factory was destroyed and three employees killed in what was reported to be a propane explosion.*

Light ends may also form explosive mixtures within process systems, storage tanks, and tank cars or truck transports. Special attention is required to remove air from a system before light ends are introduced and to prevent air from entering while light ends are present.

Major releases of LPG can be caused by:

- leakage from valve stem seals and flange gaskets;
- leakage when taking a sample or drawing water;
- leakage from transfer piping due to corrosion, mechanical damage, screwed piping connections;
- failure of transfer pipe flexible joint or cargo hose;
- leakage from a storage vessel due to corrosion;
- tank overfilling, which forces liquid out via pressure relief safety valves.

This booklet will present information on the fundamental characteristics of light ends with the main focus on LPG (see definition in Chapter 2) and 'fuel gas' as used on petrochemical plants. It will describe many tried and proven operating practices, common in the industry today, for handling light ends at refineries. Also, several important design considerations for equipment are outlined. LNG will be the subject of a future, more detailed publication.

2
Light ends defined

Light ends

Light ends, as discussed in this booklet, are either a single hydrocarbon or a hydrocarbon mixture having a Reid Vapour Pressure (RVP) of 18 psia (pounds per square inch absolute) (1.25 bars a) or more (Figure 6).

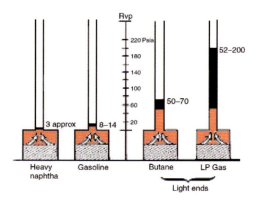

Figure 6 Hydrocarbons having a Reid vapour pressure of 18 psia (pounds per square inch absolute) (1.25 bars a) or more are light ends. Those with less are not.

Light ends will vaporize rapidly at room temperature and pressure. Common individual hydrocarbons meeting this definition are methane, ethane, propane, butane and pentane, but also propylene, ethylene, butadiene, etc. Mixtures of hydrocarbons that qualify include natural gas and fuel gas (though the latter typically includes hydrogen and other impurities). These contain a large amount of methane and ethane. LPG (liquefied petroleum gas), which is either propane or butane or a mixture of the two, is also a light end.

Refinery products (such as commercial motor gasoline and naphtha) having Reid vapour pressures less than 18 psia (1.25 bars a) are, therefore, not considered to be light-ends mixtures. However, it must be remembered that these products may contain light ends. Gasoline, for example, normally contains butane and pentane. Vapour from a gasoline spill will often contain a large amount of butane and pentane and will be hazardous.

TVP: True Vapour Pressure

The true vapour pressure of a liquid is the absolute pressure exerted by the gas produced by evaporation from a liquid when vapour and liquid are in equilibrium at the prevailing temperature and the gas/liquid ratio is effectively zero.

RVP: Reid Vapour Pressure

The vapour pressure of a liquid determined in a standard manner in the Reid apparatus at a temperature of 37.8°C and with a ratio of vapour to liquid volume of 4:1. Used for comparison purposes only.

3
ABC's of light ends

3.1 As a gas

Light ends (except C5s) are gases at atmospheric temperature and pressure. Methane and ethane (major components of natural gas) are transported long distances by pipeline. These gases are used as fuel in various industrial operations as well as for heating home and office buildings. These particular hydrocarbons, as well as other light ends, are also handled in refineries as gases in pipelines and in process equipment. Gas moving through a pipeline is normally under pressure. If a leak develops, gas will escape, mix with air and spread over a large area.

Hydrocarbon gas will burn if mixed with the proper amount of air (oxygen). Since we have no way of knowing the proportions of gas and air in the vapours around a leak, we must assume that any leak will develop into a flammable mixture. It also must be assumed that nature will supply a source of ignition. Experiments have shown that if too little gas is blended with air, the mixture will not burn. Also, if too much gas is present, the mixture will not burn. These proportions of fuel are beyond the lower and upper flammable limits. Ordinarily, we speak of a mixture being too lean or too rich to burn readily. The following table lists the flammable ranges, at normal atmospheric pressure and temperature, of several of the light-ends vapour-air mixtures:

	Volume % Vapour in Air	
	Lower Limit	Upper Limit
Methane	5.3	14.0
Ethane	3.0	12.5
Propane	2.2	9.5
Butane	1.9	8.5
Pentane	1.4	7.8
Natural gas	3.8–6.5	13.0–17.0

Many natural and fuel gases which are mostly methane and light hydrocarbons (frequently mixed with hydrogen and carbon monoxide) are lighter than air. When they leak out of any container, they rise into the atmosphere. Unless the leak is very large, these gases are diluted in a short time to the point where they will not burn. However, should the leak occur inside a pump room or other building, the gases cannot escape readily, and an explosive mixture can easily develop (Figure 7).

LNG (Liquefied Natural Gas—mainly methane) is lighter than air but when stored at cryogenic temperatures (below −162°C/−260°F), vapours coming from spilled product are very cold and dense and act like heavier gases until warming above −107°C/−160°F.

Figure 7 *Fuel gas, leaking from a ruptured 3/4-inch line, accumulated in this pump room and ignited. The resulting explosion caused the damage shown. The line ruptured when water froze in it.*

Ethane and the other light ends are much more dangerous since they are heavier than air. If these hydrocarbons leak out of a container, they will settle in a cloud along the ground. LPG vapour is twice as dense as air (Propane 1.5; Butane 2.0). They are not easily diffused into the atmosphere unless the wind velocity is at least 10 miles per hour (4.5 m/s) (Figure 2). Also, at ground level there are usually many more sources of ignition, such as internal combustion engines, welding equipment and fired heaters (Figure 8).

Figure 8 Some ignition sources.

3.2 As a liquid

Light ends are often stored and handled as liquids. This is particularly true of propane, butane, pentane and LPG under pressure, and methane and ethane refrigerated.

Pressure

To keep propane as a liquid at 100°F (38°C), it must be kept at a pressure of at least 189 psia (13 bars a); normal butane, 52 psia (3.6 bars a); isopentane, 21 psia (1.5 bars a); and LPG, 52 psia (3.6 bars a) to 189 psia (13 bars a). If these pressures are not maintained, the liquid will vaporize quickly. Thus, a small amount of liquid leaking from equipment handling light ends will vaporize and form a large blanket of vapour (Figure 2). This is why it is so important to keep light ends confined.

Heavier hydrocarbons at atmospheric temperature do not have to be handled under pressure and, normally, do not vaporize sufficiently to be a hazard except in the immediate area of a leak.

Viscosity

The relatively low viscosity of light ends compared to heavier hydrocarbons adds to the problem of containing them within pressurized equipment (Figure 1). At 100°F (38°C), kerosene is about fifteen times as viscous (thick) as propane and six times as viscous as pentane; thus, light ends flow through flanged joints and packing much more easily than heavier hydrocarbons. Light ends will flow through openings where even water cannot go.

Boiling point

The low boiling points at sea-level atmospheric pressure of propane (−43.8°F/−42°C) and butane (31.1°F/−0.5°C) create hazards in depressuring equipment in which there may be water or heavy hydrocarbons. As water freezes, it expands, applying tremendous pressure on the equipment (refer to BP Process Safety Booklet *Hazards of Water* for more details). Also, equipment metallurgy must be adequate to prevent brittle fracture at low temperatures.

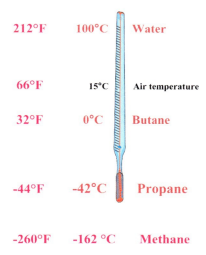

ACCIDENT A combination cracking unit was severely damaged by a fire and explosion, and one fatality resulted, because it was not recognized that vaporizing butane could lower the surrounding temperature below that required to freeze water. In this accident, a stabilizer tower and reflux drum were being pumped out, depressured and purged of hydrocarbons in preparation for a shutdown. During depressuring, residual light hydrocarbon vaporized, cooling the drum and freezing water in a drain connection. Since nothing could flow from the drain, the operators assumed all liquid was out of the drum and left the drain valve open by mistake.

Later, the drain thawed, and a large amount of liquid escaped and vaporized. The gas flowed along the ground, blanketing a large area before it flashed at a furnace. Had the operators recognized that the drain could be blocked with ice, they would have made sure it was open (Figure 9), found that the drum was not empty, closed the drain valve and proceeded to remove the hydrocarbons from the drum safely. Another lesson to be learned from this accident is that equipment should be thoroughly steam purged, if possible, to be certain that all residual light ends are vaporized.

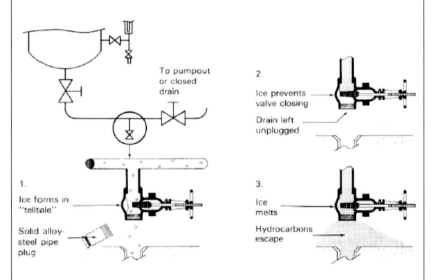

Figure 9 Make sure that drains are not iced.

Water drain procedures are given later in this booklet (refer to Section 8.2).

ACCIDENT A flash fire followed by a BLEVE (Boiling Liquid Expanding Vapour Explosion) occurred within the LPG tank farm area of a refinery at Feyzin, near Lyon, France. Eighteen people were killed and 81 injured. The gas storage facility, which contained four 1200 m³ propane and four 2000 m³ butane storage spheres, was destroyed. Two spheres suffered BLEVE's, creating a crater 35 × 16 metres × 2 metres deep (115 × 52 × 7 ft). A piece of sphere weighing 63.5 kg (140 lbs) was found 300 metres (1000 ft) away. The fire spread to nearby liquid hydrocarbon storage tanks.

The events leading to the accident started with a water draining operation on a propane storage sphere. Water originating from process treatment units accumulated in the bottom of the spheres and had to be periodically drained off. The water draining arrangement consisted of an uninsulated pipe that discharged to ground underneath the sphere. Both valves had been opened. When the operation was almost complete, the upper valve was closed and then cracked open again. Initially nothing came out of the pipe, so the operator continued to open the upper valve to clear out what he assumed to be ice or hydrate when the blockage suddenly cleared with a large release of liquid propane. The operator found it impossible to close the upper valve again as it had frozen in the open position due to the low temperature created by the liquid propane flashing to vapour across the valve (−44°F/−42°C). He then tried to close the lower valve, but this too was frozen in the open position. With a clear path to atmosphere, liquid propane continued to escape forming a vapour cloud that drifted across the site, through the boundary fence across a motorway, 60 metres (200 ft) away. Although traffic was stopped on the motorway, the gas cloud was ignited by a passing car about 160 metres (525 ft) away.

The sphere was engulfed by a large fire fed by the escaping propane that caused the pressure relief valve to lift. Escaping propane vapour ignited. The local fire brigade arrived quickly, but they were untrained in how to fight an LPG sphere fire. They concentrated their efforts on cooling the adjacent spheres. After 1 hour, the burning sphere exploded releasing a wave of liquid propane that burned in a rising column of fire giving out very high levels of thermal radiation (BLEVE). The firemen nearby were killed. Fragments of the sphere cut into the legs of an adjacent sphere, causing it to topple over, with its relief valve discharging liquid propane into the fire. Half an hour later a second sphere suffered a BLEVE. Three other storage spheres collapsed and ruptured, as their supporting legs were not provided with any fire protection.

The low boiling points of butane and propane create two other operating hazards. One of these is that operators may suffer from severe frostbite if liquid butane or propane contacts their skin.

The rapid vaporization of these liquids can cool the skin sufficiently to cause frostbite. Gloves and goggles should be worn if there is danger of exposure of hands or eyes to liquid propane or butane. If butane or propane contacts your skin, remove the liquid immediately and wash the affected area with lukewarm water.

The other hazard is that liquid butane and propane vaporize when suddenly released and frequently have the appearance of a white steam cloud. When a cloud is observed in an area where light ends are processed, it should be viewed with suspicion. Investigate immediately to be certain that it is steam and not a dangerous cloud of hydrocarbon.

The white cloud which sometimes appears around a liquid-propane leak results because the vaporizing liquid absorbs heat from the atmosphere and condenses particles of moisture in the air. However, this does not always happen, and you cannot depend on being able to detect a light-ends leak by looking for a white cloud.

Expansion

Almost all liquids and solids will expand as they become warmer. This is why the liquid in a thermometer rises as it becomes warmer. Expansion due to an increase in temperature is called *thermal expansion*.

If for example, one litre of propane at 0°C is heated to 40°C its volume increases to about 1.13 litres. The same volume of butane subjected to the same temperature rise expands to about 1.08 litres. In general, the lower the density the greater the rate of expansion or contraction. This coefficient of expansion is about 10 times that of water, and must obviously be taken into account when a container is filled with LPG. If a container is filled completely with LPG and then subjected to a temperature rise, the developed pressure rapidly rises above the design pressure of the container, because the expansion is constrained. To avoid the risk of this, care is always taken to ensure that a vapour space remains in all operating conditions. Limits are defined in design codes and local regulations.

Thermal expansion of light hydrocarbon liquids can cause serious problems. Two failures of sample containers in refinery laboratories were due to overpressure because of expansion of the liquid as it warmed up, and they could have been avoided if the potential danger had been recognized and certain precautionary measures taken.

To minimize this hazard, the following general rules for the safe sampling of light liquid hydrocarbons were developed. These rules apply to sampling procedures for all liquid hydrocarbons having a Reid vapour pressure of 18 psia (1.25 bars a) or more.

ACCIDENT The first of these failures was a 1,900 millilitre sample container that had been filled with field butane at a line pressure of 220 psi (15 bars) and a temperature of 26°F (−3°C) (Figure 10a). This was an Air Force surplus oxygen container of two-piece welded steel construction with a design pressure rating of 500 psi (34.5 bars) and a working pressure rating of 400 psi (27.6 bars). The regular sampling procedure called for venting the container to assure an adequate vapour space, and this step apparently was overlooked. The container was brought into the laboratory liquid full, without any contained air dissolved in the butane.

In the laboratory, the contents of the container approached the room temperature of 75°F (24°C). The resulting thermal expansion of the liquid could have subjected the container to a pressure of more than 1,000 psi (69 bars), if bursting had not occurred. When it burst, the vaporized contents flashed at a furnace in which carbon was being burned from a sample of cracking catalyst.

Figure 10a A 1,900-millilitre sample container ruptured, and the vapour flashed. Thermal expansion of the liquid resulting from a 49°F (27°C) temperature rise could have subjected the container to at least 1,000-psi (69 bars) pressure, if bursting had not occurred.

The sample containers are to be of 75 and 150-millilitre capacity and are to be manufactured of monel metal. Under special circumstances, containers as large as 300-millilitre nominal capacity may be used, provided that special permission is obtained from an authority designated by the refinery manager.

Sample containers are to be designed for a pressure rating of 5,000 psi (345 bars) and hydrostatically tested at 8,000 psi (552 bars). A hydrostatic test of 8,000 psi (552 bars) is to be performed once every five years, and the containers are to be inspected visually at frequent intervals for corrosion. Figure 10b compares the Air Force surplus type of container with the 5,000-psi (345 bars) unit.

SAFE HANDLING OF LIGHT ENDS

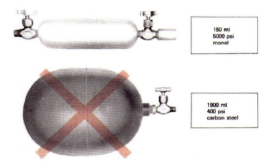

Figure 10b Comparative data for a surplus Air Force oxygen container (similar to Figure 10) and a sample container rated at 500 psi for light liquid hydrocarbon service.

Samples should be taken by qualified personnel only. The samples should always be taken with the knowledge of the operator responsible for the unit.

The following are important safety measures.

- After a container is filled, liquid must be withdrawn to provide at least 20% of the container volume for vapour space for possible expansion. Valves and piping can be arranged to give this desired outage.
- To avoid overheating, filled containers should be kept out of doors in the shade or under refrigeration.
- Sample containers must be prepared prior to sampling so as to avoid explosive mixtures.
- Use the proper sampling container (Figure 11).

The sample containers normally employed are of the fixed volume type, and may be provided either with a single valve or an inlet and outlet valve in addition to an ullage tube that is used to ensure that filling does not go beyond the 80% full limit. Typical examples are shown below.

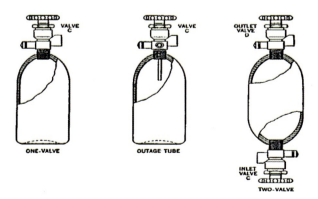

Figure 11 Sampling containers.

Pressure-relieving devices are not used on sample containers even though they would prevent rupturing from excessive pressure. Such devices would release light hydrocarbon in a laboratory. This could be very hazardous.

Density

Liquid LPG has half the density of water (LNG 0.45; Propane: 0.50; Butane: 0.57), so a liquid spill will float on water.

Heavy hydrocarbons can act like light ends

It is extremely important to remember that hydrocarbons which are heavier than light ends behave similarly to light ends when they are handled at high temperatures. Hot releases of heavy hydrocarbons can auto-ignite if they are hot enough.

At high temperatures, these hydrocarbons must be held under pressure to keep them liquid; their viscosity is lowered, and they vaporize readily if the pressure is removed (Figure 12). Operators on all refinery process units must never forget this.

Released vapours can cool and form aerosols that will ignite way below that apparent flash point of the parent liquid (refer to BP Process Safety Booklet *Hazards of Air and Oxygen* for more detail on this and on auto-ignition temperature).

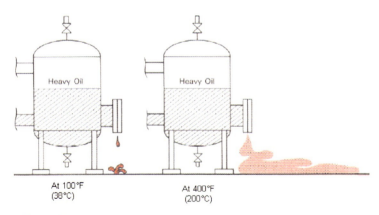

Figure 12 Heavy hydrocarbons at high temperatures act like light ends.

Toxicity

Most light ends are not acutely toxic, however, some (like butadiene) may have long term health effects—check the relevant Material Safety Data Sheets.

All light ends possess mildly anaesthetic properties and are asphyxiant as they can deplete a confined space of oxygen by pushing it out.

- Make sure that no LPG containers are stored in enclosed spaces, including laboratories. Use outside well-ventilated areas, in shade and away from drains, doors and windows for storage.
- All rooms where piping is likely to leak (such as from flanges, valves) should be considered as confined spaces.
- Beware of low points (for example, sewers, drains, pits, trenches) where LPG can accumulate.

For more detail on asphyxiants, refer to BP Process Safety Booklet *Hazards of Nitrogen and Catalyst Handling*.

Colour and odour

Light ends are colourless in both the liquid and vapour forms. Any coloration of the liquid observed, say during sampling, should be investigated immediately, as it could indicate contamination.

The cloud that appears when LNG or LPG leaks from any point is white, but this is not the colour of the product itself. It is chilled water vapour condensed from the air by the evaporating LNG or LPG. It cannot be trusted to show the limits of the flammable cloud (LEL-UEL) but is a good indicator of the gas direction.

LPG for commercial use must be given a distinctive odour so that it can be detected as soon as a leak occurs. The British Standard Specification BS 4250:1987 Parts 1 and 2 requires that the gas be detectable in air at concentrations of one fifth of its lower limit of flammability. Odorants in use throughout the industry include ethyl mercaptan, thiophane and amyl mercaptan.

Unsaturated LPGs (like propylene, butylenes, etc.) do smell strongly.

Other hazards

LNG is stored as a liquid in refrigerated vessels below $-162°C$ ($260°F$). Any contact with a liquid spill can result in severe cold burns (instant frostbite) and death if the affected surface is significant.

4
Light ends are everywhere

We must recognize that light ends are found nearly everywhere in refineries.

Process units involved in light-oils operations (such as alkylation, isomerization, polymerization, reforming, gasoline blending, catalytic cracking, crude topping and several others) all handle light ends.

LPG products are loaded, unloaded and stored in refineries. Propane storage drums, butane spheres, pentane spheriods and fuel-gas holders may be found at many locations throughout refinery process areas and tank fields.

Light ends also are found on many heavy-oil process units. Propane is stored and mixed directly with heavy oils in propane dewaxing units and in propane deasphalting units. Some solvent dewaxing units, such as MEK units, also utilize propane, but only as a refrigerating agent.

Remember that everywhere fuel gas is handled, light ends are present.

5

Handling light ends on process units

Air and light ends must not be brought together in refining equipment except under rigorously controlled conditions. If air and hydrocarbons are permitted to mix in flammable proportions, it can be assumed that nature will provide a source of ignition which will start a fire.

Operators should remember that in many refinery process units, the light ends may be processed at temperatures high enough (750°F/400°C and higher) to self-ignite in air. This temperature is called the *autoignition (spontaneous-ignition) temperature*.

Operators must always be on the alert for leaks of any kind. Pump and valve packing should be watched carefully for leaks and if any occur, the packing glands should be pulled up. Drain and bleed valves must be kept closed and plugged and the action logged (Figure 13).

Solid alloy steel plugs

Figure 13 *When drainage is complete, drains should be plugged or blinded. These drain points should be checked regularly to confirm plugs/blanks presence.*

5.1 During start-up

Air must be eliminated from process equipment before light ends are introduced on start-up. The techniques for accomplishing this are described in the BP Process Safety Booklet *Safe Ups and Downs for Process Units*. One important fact to remember is that just one gallon (3.8 litres) of liquid propane introduced into a 10,000 gallon (38 m^3) vessel containing air at atmospheric pressure and at 60°F (15.5°C) will provide enough fuel for a serious explosion (Figure 14).

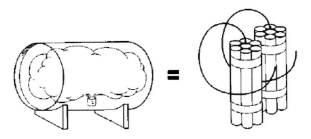

Figure 14 One gallon of propane vaporized in a 10,000-gallon (38 m^3) air filled drum has the energy equivalent of 13 pounds of TNT. A source of ignition can turn that innocent looking drum into an exploding bomb.

Water must also be eliminated on start-up. The start-up procedure of most units requires steam purging through vessels, drums and lines to purge air from the system and, in some cases, requires steam testing of vessels to detect leaks. During the steaming period, water will collect in pump casings, bottoms of vessels, and low spots in lines and exchangers. This water must be drained frequently.

There is little, if any, danger of tank foamovers where light ends are handled since they are not normally stored at temperatures over 200°F (93°C), and they vaporize before water begins to boil.

However, a problem can arise when water is not drained from a light-ends storage tank that provides charge stock for a processing unit. If such charge stock is pumped to a fractionating tower, the water may flash to steam in the tower and cause serious damage to the internals.

Further details about the hazards of water in process units may be found in the BP Process Safety Booklet *Hazards of Water*.

5.2 During shutdown

Residual light ends must be removed from refining equipment before air is admitted. This is accomplished by a steam or other inert-gas purge, followed whenever it is possible by overflowing the equipment with water.

Detailed techniques on how this should be accomplished are outlined in the BP Process Safety Booklet *Safe Ups and Downs for Process Units*. The low boiling points of propane and butane create hazards in depressuring equipment containing water or heavy hydrocarbons. When propane and butane vaporize, temperatures can be lowered sufficiently to freeze water in drains. At one refinery, the operator's failure to recognize this resulted in a serious fire.

Details of this problem are outlined in Chapter 3.

Blinds must be installed as close as possible to the closed valves used to exclude light ends. By locating blinds near the valves, the volume of light ends released from between the blind and valve, when the blind is removed, is minimized. This principle holds for all oils. It is particularly important with light ends since they leak past valves readily because of their low viscosity, and they vaporize quickly.

Maintenance and turnaround items

The (de)commissioning procedures should take account of the following checks and inspections:

Pressure relief valves

Pressure relief valves should always be removed, tested and inspected and re-set at frequent intervals, based on local regulations and previous history; and a written scheme of examination (WSE) drawn up by a competent person.

A register of relief valves should be maintained and each individual valve identified by a works identification number.

The relief valve tail pipe should be adequately supported and covered with loose fitting plastic caps to prevent ingress of water.

The tail pipe should have a drainpipe provided which directs away from the vessel to prevent any flame impingement on to the shell.

Vessels

LPG vessels should be inspected internally at least every 10 years and hydrotested or in accordance with the WSE (specific inspection techniques may be applied to replace or complement the hydrotest).

Welds

Radiography or other non-destructive (or non-invasive) inspection method of piping welds should be carried out regularly as per inspection schedule.

Water accumulation

Nozzles and areas where water may accumulate and cause pitting corrosion should be subjected to more frequent inspections involving both external examination and ultrasonic testing.

Ground slope/gradient

Check after vessel overhaul or other major works in the bund to ensure slope away from under sphere or bullet has not been changed or altered by general maintenance work, civil works on the bund floor or vegetation growth or other event that may prevent liquid LPG draining down the slope away from under the vessel.

Filters/strainers

Checks should be carried out for pressure drops every six months.

Stairways/handrails

Stairways and handrails and their supports should be checked for signs of corrosion and for general wear and tear.

Pipe supports

LPG piping supports should be checked for alignment, function and resting positions to ensure piping is adequately supported.

Passive Fire Protection (PFP)

Regular checks must be made to ensure that PFP integrity is intact on vessel, supports, isolation valves, etc.

5.3 During processing

Light ends often come from processes where they are mixed with for example, H_2S, cyanides, phenols, ammonia. This may create metallurgical/operational issues and not dealing with these properly has led to many major accidents.

Operating procedures

LPG facility operating manuals should be readily available at the control room location. This manual should include emergency instructions, maintenance procedures, Safety Checklist, flowsheets and process and instrument diagrams—all of which must be up to date.

Facility technicians' or operators' knowledge and skills for routine and emergency operations should be checked at annual intervals to ensure competence.

Air hazards

Operators must be constantly alert for the entry of air (oxygen) into process equipment, especially so when changes in operating methods are made on a particular process unit or on units supplying feed for it. Air or oxygen may enter light-ends facilities when (1) dissolved oxygen is present in the feed stream, (2) equipment is operated under a vacuum, and (3) water is used for process washing. Also, it is possible that air may be accidentally introduced into processing equipment.

For example, injection systems used for adding corrosion inhibitors or antifoam agents must not be pumped empty. In such an event, air could enter the injection line, or hydrocarbons could back out to the atmosphere through the same line. *Remember, hydrocarbons must not be mixed with air in pressure vessels or lines except under rigorously controlled conditions.*

ACCIDENT Propane storage drums and an alkylation unit (Figure 15) were extensively damaged (Figure 16) because of the unrecognized introduction of oxygen. In this case, oxygen was carried in solution with the propane stream.

Figure 15 Arrows indicate location of three propane storage drums near an alkylation unit.

Figure 16 Propane enriched with oxygen detonated. Ignition was probably caused by iron sulphide. The three propane storage drums (Figure 15) fragmented. Arrow (Figure 16) indicates remains of one of the propane storage drums. Other equipment was damaged by fire from the released hydrocarbons.

Laboratory research has shown that oxygen may be absorbed by oils and later released (somewhat like bubbles in soda water). This may cause a nonflammable vapour space to become flammable by the process of oxygen concentration or enrichment. Let us see how this is possible.

Assume that the propane stream contains a small amount of oxygen dissolved in the liquid. When the propane is pumped into storage, some propane vaporizes into the space above the liquid. Also, along with the propane vapour, oxygen leaves the liquid, and an equilibrium vapour phase is built up containing a much higher concentration of oxygen than was present in the liquid (Figure 17).

Now suppose that the liquid level is raised in the vessel and the vapour space is not vented. Much of the propane vapour under pressure condenses back to a liquid; but the gaseous oxygen, because it redissolves very slowly in the liquid, becomes more concentrated in the vapour because of compression (Figure 18). The percentage of oxygen in the vapour has thus increased many times over what it was in the liquid entering the storage vessel. Lighter hydrocarbons, such as propane and butane, are particularly hazardous. They can carry with them more oxygen than heavier hydrocarbons, and they frequently are stored under conditions which can lead to the oxygen concentration effect described.

When there is a possibility of air accumulation anywhere in a process unit, the vapour spaces of the equipment involved should be sampled routinely, and routine checks for oxygen in the feed should be considered. A value of 0.4% oxygen is often used as the maximum admissible value before loading LPG.

Figure 17 Equilibrium vapour phase.

Figure 18 Propane condenses more readily into liquid under pressure, leaving the vapour space more concentrated in oxygen.

The sampling of liquid light ends requires much care. General rules for the proper sample container to use and general procedures for obtaining samples are described in Section 3.2 under *Expansion*.

Hot work

Hot work, such as welding and cutting on an operating unit, should not be permitted inside the unit limits except under rigorously controlled conditions—likewise, vehicles powered by internal combustion engines or other drives that provide a source of ignition should not be allowed within battery limits of an operating unit without written authorization of that process unit.

Occasionally, hot work is permitted close to areas where light ends are present. However, this is done only when operating management has made certain that the operation will be safe and will continue to be safe until completed. The determination of the safe operation requires testing of the atmosphere for combustibles at the hot work location and at the adjacent areas, under strict adherence to good 'permit to work' rules.

This can be done with combustible-gas analyzers. Several instruments are available, both manual and automatic (continuous) types.

More information on this subject is presented in the BP Process Safety Booklets *Hazards of Air and Oxygen, Confined Space Entry* and *Safe Tank Farm and (Un)loading Operations*.

Water hazards

Operators must be particularly careful when draining water to open drains from vessels containing light ends, and the operator must be present during the entire operation. Operators must be certain that they have closed the valves tightly and have replaced the plugs after draining water. Ice in a valve will also stop flow and prevent complete closing of the valve. When the ice thaws later, hydrocarbons will escape (see Figure 9 on page 10).

Water drawoff, whether manual or automatic, should be to a closed system. An open drain or 'telltale' should be installed in the drain line to determine when the water has been completely drained to the closed drain.

In some instances, it may not be advisable to remove all of the water, with resultant light ends 'blow through'. In this event, a small water collection pot can be installed with a gauge glass to permit controlled water drawing with a water seal at all times. This installation must meet the process system specifications and must have freeze protection if needed.

Water used for process washing or for flushing can carry air into hydrocarbon systems. This air may accumulate in a process unit, developing into a flammable mixture of hydrocarbon and air. Whenever water is used for process washing or for flushing, vapour spaces of the process equipment involved should be sampled routinely for oxygen.

Gasoline-blending hazards

The blending of butane into gasoline base stock in atmospheric storage must be avoided. A closed circulating pipe system (under pressure) should be used.

When butane is introduced into a tank containing a low-vapour-pressure stock, the initial charge of butane may cause vapour agitation of the stock, especially if the liquid level is low. Such agitation may generate electrostatic sparks on the oil surface and result in a serious explosion before the oil surface is blanketed by the butane vapour.

This operation is believed to have caused a refinery tank explosion and fire involving about a $1,700,000 loss (Figure 19). Only a small amount of butane had been added to the tank when the explosion occurred. Static electricity was believed to have been the source of ignition.

ACCIDENT

Figure 19a Tank blending of butane is thought to have caused this tank farm fire.

An explosion occurred in the tank while butane was being blended with gasoline. The roof was blown off. It fell on pipe manifold, broke cast-iron fittings and started a ground fire.

The incident above occurred in the 1960s. The one below occurred in the 21st Century, showing that some lessons from the past are still valid.

ACCIDENT Batch blending was going on in a 7,000 m³ unleaded gasoline tank when a fire occurred. Fifty-six fire trucks tackled the fire, over more than 30 hours, successfully protecting adjacent tanks. The investigation team found that the blending calculations were wrong, and three times too much butane was being sent to the tank. A bubble of light ends probably lifted and tilted the roof, creating enough static or metal to metal friction to ignite the vapours.

Figures 19b & c

Bad control of blending operations or of high RVP products has caused multiple floating roof sinkings.

Relief valves

Relief valves are installed on equipment to prevent internal pressure from building up to a point which might damage the equipment. Special facilities operated by trained personnel are available to sites for testing and resetting relief valves to the proper pressure. No one else is to tighten down or otherwise alter in any manner the setting of a relief valve.

Flare-stack hazards

A flare-stack system may become a fire and explosion hazard unless given proper attention. The stack flame, pilot light and lighting device can be sources of ignition. Also, the stack knockout drum or piping may contain pyrophoric iron sulphide deposits which catch fire spontaneously when exposed to air. A fire or explosion could result if air were leaking into a flare-stack system.

Flare systems are usually designed with a continuous purge or water seal to prevent air from being drawn back into the system.

On catalytic reforming units, the mixture of flared gas is often lighter than air, largely because it contains appreciable quantities of hydrogen. This lighter than air gas may mean that a vacuum will be present in the piping and knockout drum at the base of the flare.

Under these conditions, air will try to get into the system. Operators must be careful to see that no bleeders or vents are left open.

On starting up, air should be purged from a flare system with steam or other inert gas before hydrocarbons are vented.

Some ignition systems for flare stacks have air and fuel gas connected to a common header. Wherever such systems are in use, the air and fuel-gas connections should be disconnected or blinded after the flare is ignited. This prevents air from leaking into the fuel gas or into the flare stack.

Drain the flare-stack knockout drums regularly so that slugs of liquid are not vented out the stack. If automatic liquid level controllers are used, check their operation often. Any liquid vented will be ignited by the flare and fall to the ground where it may cause a serious fire or injury (Figure 20). A flare was pumped over at one refinery, and the resulting fire caused extensive damage.

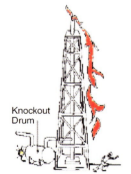

Figure 20 *Burning liquid may be vented from flares if the knockout drums are overfilled.*

Also, flare stacks have frozen on refrigerated storages because steam was supplied to the flare tip and condensate fell back into the stack.

Refer to Appendix 1 for regular checks to be conducted during routine operation of LPG facilities.

6
Storage of light ends

Light ends are stored underground or above ground either as liquids or as gases, depending primarily on the particular hydrocarbon or hydrocarbons involved. In either case, they may be stored at ambient temperatures or under refrigeration. Except for a brief description of underground storage, this booklet will be limited to the discussion of above ground storage at atmospheric temperatures.

The various types of underground storage include:

- washed-out salt caverns;
- hardrock (mined) caverns;
- depleted gas wells;
- in-the-ground surface pits; and
- high-pressure storage in buried pipe cylinders.

The use of underground storage has increased in recent years because of the low construction and maintenance costs, the conservation of real estate and, particularly because it is safe.

In above ground storage, methane and ethane are stored as gases either at atmospheric pressure in gas holders (Figure 21) or under pressure in pressure vessels (Figure 22). Propane and butane are maintained as liquids in pressure vessels such as spheres (Figure 23a). Underground storages have become common containers for LPG (Figure 23c). Isopentane generally is stored as a liquid in spheroids (Figure 24). LPG are also sometimes stored in refrigerated vessels (Figure 23b).

Non-flammable atmospheres must be maintained within storage vessels, and lint ends must be confined at all times. Let us examine some of the important procedures to accomplish this.

Figure 21 Gas holder.

Figure 22 Pressure vessels.

Figure 23a Spheres.

Figure 23b Refrigerated storages. There are two types of cryogenic tanks, metal double wall and PS (Pre-Stressed) concrete type.

Figure 23c Underground storage.

Figure 24 Spheroid.

6.1 Activating storage vessels

Removing air from the lines and vessels before hydrocarbons are introduced is the principal problem in this operation.

Before closing the vessel, an inspection should be made to assure that no foreign materials remain and that the interior is clean. Manways and pipe flanges are then immediately bolted up to prevent foreign material from entering to damage or foul the equipment (Figure 25). This inspection should include a check of the closure to assure that gaskets are properly placed and all bolts are properly tightened.

Figure 25 Close equipment immediately after inspection to keep foreign material out.

Hydrostatic testing may be carried out on vessels designed for it. If so, gauging devices and low and high level alarms should be checked for reliability while filling and emptying.

Gas holders should be filled with inert gas before hydrocarbons are introduced to make certain that they will rise and fall properly and that the instruments provided are functioning correctly.

A blind list, prepared before deactivation, will show each type of blind, its number, size, location, date installed, the initials of the person who installed it, and will also provide spaces in which to indicate the date that each blind is removed and the initials of the person who removed it. Each blind location may be assigned a permanent number. The blind list must be followed to make sure that all blinds requiring removal in the first step of activation have been removed as required at the proper times during the activation (Figure 26). If blinds are reinstalled after being removed and signed out, this must be noted on the blind list. No blind may be installed or removed without notice to and permission from the operators. All disturbed joints must be identified.

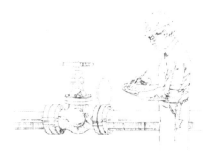

Figure 26 Records are important to make sure blinds are installed and removed.

Air may be removed through a top vent connection by purging with steam or other inert gas. Care must be taken to be certain that air is removed from all lines to and from the storage vessel. Blinds should be removed in accordance with the specific instructions.

A device for determining the air content of the steam purge is illustrated in Figure 27. If tests show that oxygen is still present in a concentration greater than 1.0 percent by volume (equivalent to 5.0 percent air), purging must be continued.

Figure 27 This is a device for determining the amount of air, or other noncondensibles, present in steam. A representative portion of the steam flows through the sample chamber, heating the sample chamber to steam temperature. Placing the sample chamber in the insulated container speeds up this step. A sample of the steam is then trapped in the chamber by closing in quick succession the inlet valve and outlet valve. The chamber is then lifted from the insulated container and cooled to 100°F (38°C). Condensation of the steam produces a vacuum in the chamber. The amount of pressure reduction is a function of the relative proportions of steam and noncondensibles (air) in the original mixture, and the air content can be calculated. This instrument cannot be used to determine concentration of air in inert gas. After purging with inert gas, absence of oxygen must be confirmed using one of several types of portable oxygen analyzers suitable for this purpose. Drawings of this analyzer can be found by contacting Amoco heritage refineries.

If nitrogen or gas from an inert-gas generator is used to purge the system instead of steam or water, the same general procedure should be followed. The main difference will be that drains and vents will be kept open only long enough to ensure complete drainage of any liquids present and until analysis by Orsat or another method, such as a portable oxygen analyzer, shows the oxygen content of the vented gas to be no more than 1.0 percent.

When possible, it is preferable to fill the storage vessel with water to push the air out. However, before filling a storage vessel with water and, in particular, before filling any sphere or spheroid with water, check with the engineering department to be sure that the vessel, its foundation and supporting structure were designed and are still in a condition to support the water load; otherwise, the vessel may rupture, settle or collapse (see BP Process Safety Booklet *Hazards of Water*). After filling, the water is backed out to the sewer with fuel gas, nitrogen or vapour from another storage vessel which contains stock similar to that to be stored.

6.2 De-activating storage vessels

One of the first steps in de-activating or emptying a storage vessel is to install blinds in the feed or inlet lines to ensure that no fresh hydrocarbon can enter the vessel.

Any stored liquid must be pumped to a prescribed place. Centrifugal pumps should be watched closely to make certain that none completely loses suction before it is shut down. Running a centrifugal pump dry for even a short time may seriously damage it.

Reciprocating pumps are best for pumpout because they have superior suction characteristics so are less susceptible to damage on loss of suction. Butane and propane, because of their high volatility, are especially difficult to pump out when liquid levels become very low, and it may be necessary to install temporarily a portable reciprocating pump under a storage vessel to pump it out. Some operators pump a cold light oil into butane spheres when the level gets low in order to absorb the butane and to obtain better suction conditions for their pumpout pump. If blinds in the feed line must be removed to add oil, great care must be used to prevent fresh butane from entering.

The next step is to blind off the pumpout lines and vent lines connected to other storage vessels still in service. The remaining hydrocarbons should then be removed by purging the vessel with steam or other inert gas out through a top vent to the flare, then finally to the atmosphere. Purging with steam is preferred where heavy hydrocarbons may have entered the vessel, as may happen when LPG gas has been transported through a crude pipeline. Continue purging until the combustibles coming off the vents and drains are below 1.0 percent.

Note that some sites use internal combustion engines as in many cases the storage area is not connected to a flare. This is especially important with carcinogens such as butadiene.

If steam is used for purging hydrocarbons from a light-ends storage vessel, the air-in-steam analyzer shown in Figure 32 can be used to indicate the light hydrocarbons present as noncondensibles. This technique is not accurate to more than 1 or 2 percent at the low concentrations required for safety. Therefore, depending on the individual situation, operating procedures should specify some amount of additional purging after a safe condition is indicated by the instrument.

The hydrocarbon content of issuing purge gas cannot be measured directly with the normal combustible gas tester, as this type of instrument requires air to burn the combustibles in the instrument (refer to BP Process Safety Booklet Hazards of Nitrogen and Catalyst Handling *for more details).*

Specific instruments, such as InfraRed (IR) detectors are available for measuring the hydrocarbon content of nitrogen or other noncondensible purge gases.

Figure 28 Combustible gas indicators operate on either the combustion principle, the thermal conductivity principle, or on the infrared principle. To read percent hydrocarbon directly, they must be calibrated for the gases involved, and the proper scale must be used.

The portion of lines (fill, suction, etc.) between the blind and the vessel also must be purged. The purging process may require several hours, the actual time depending on the storage volume.

It is sometimes falsely believed that purging of gas holders is very simple and that no hazard exists since there is very little gas space when the holder is down. However, gas holders must be purged carefully.

ACCIDENT On November 14, 1927, a large gas holder in Pittsburgh, which was believed to be thoroughly purged of gas, was entirely wrecked (Figure 29). Thirteen men at work repairing the tank were killed, as were 15 other persons nearby. Ten million dollars was the estimated property loss.

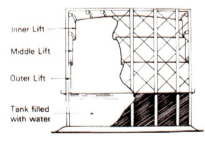
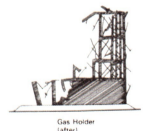

Figure 29 Only one quarter of this gas holder frame was left after it exploded.

Pressure storage vessels, which can withstand the hydrostatic pressure and support the weight, should be filled with water and overflowed. The importance of venting off gas at the highest points of all pockets in the system cannot be overemphasized. Due consideration must be given to internal partitions, stiffeners, nozzles, etc that could trap oil and gas.

ACCIDENT A 45,000-barrel (7,200 m^3) noded spheroid (Figure 32) was ruptured by an internal explosion as purging water was being drained. A wall of water leaped the fire wall and damaged a 118,000-barrel (18,800 m^3) floating-roof tank (see arrow in Figure 31b). The spheroid was being purged preparatory to inspection. Oil was pumped out, leaving not more than 25 barrels (4 m^3). Then the spheroid was water filled (46-foot (14 m) gauge) and overflowed through the overhead vent (51-foot (15.5 m) level) until no oil was found in the effluent. The top manway cover was next removed and drainage started. On removal of the manway cover, a black scum about 1/2-inch thick was noted, and bubbles were seen breaking through the scum. An odour was evident, but no gas test was made. Draining continued for almost ten hours, lowering the water to about the 37-foot (11.3 m) level. The internal explosion caused a split which extended 60 feet (18 m) horizontally and 32 feet (10 m) downward.

Investigation and sampling indicated that the probable source of ignition was iron sulphide. Air entered the spheroid through the open top manway. Fuel was provided by the trapped oil and vapour in the dome and nodes shown in Figure 32. After flooding a vessel, the water should be drained to the sewer. Sufficient vents and manways should be opened so that outside air may flow in and prevent the pulling of enough vacuum to collapse the storage vessel.

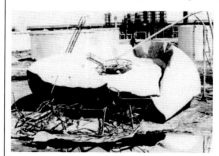

Figure 31a Spheroid rupture was caused by low-grade explosion. Vacuum damage was due to an outrush of water following explosion.

Figure 31b Water wave from collapsed spheroid overflowed the fire wall and damaged a tank in the background.

continued

SAFE HANDLING OF LIGHT ENDS

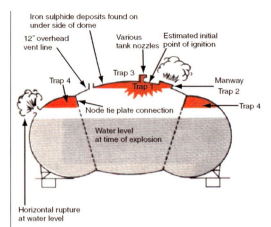

Figure 32 A schematic cross section of the noded-type spheroid showing traps for oil and vapour. Trap 1 existed because the overhead vent nozzle was not located at the topmost point of the spheroid. Traps 2 and 3 were occupied by vapour and oil confined in the manway and other nozzles. Trap 4 was a pocket running entirely around the spheroid in the space enclosed by the node tie-plate connection, the spheroid shell and the rising water.

The design made it impossible to eliminate all vapour and oil from the spheroid by overflowing with water, and it violates the principle that purging cannot be satisfactorily unless vents are at the high points of the vessels and traps are avoided. Had proper connections been provided at high points in each trap, all oil and vapour could have been removed from the spheroid by overflowing with water until clear water flowed from each overflow connection.

Recently, pressured storage vessels suspected of containing pyrophoric iron sulphide deposits have been acid cleaned prior to opening. This technique has proven very successful in eliminating the autoignition potential of these deposits, particularly where these were difficult to get to in order to keep wetted. When opening a storage tank in which a stock containing sulphur has been stored, deposits of iron sulphide may be found. These deposits, in the presence of air, will ignite as soon as they dry, so they must be kept wet until they are removed (Figure 33). The simple burning of this material alone may do considerable damage to a storage tank. Refer to BP Process Safety Booklet *Safe Ups and Downs for Process Units* for more details on pyrophoric scale hazards.

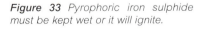
Figure 33 Pyrophoric iron sulphide must be kept wet or it will ignite.

Note that other contaminants such as HF, nitriles, etc. may be present depending on the upstream process.

6.3 Operations during active service

Water must be drained regularly. This is especially important when the stored material is pumped to process units as feedstock, and in fuel-gas storage where drops of water or condensed hydrocarbons from a storage drum or knockout drum could extinguish a furnace burner flame (Figure 34). When drawing water the precautions described in Section 5.3 *Water hazards* should be taken.

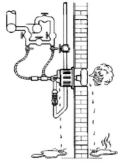

Figure 34 Slugs of liquid from the gas line may put out fires.

One type of fuel-gas holder utilizes a water seal. Condensed hydrocarbon and entrained oil in the fuel gas tend to collect on the water surface. This oil must be skimmed off regularly. The water should be continuously replaced to avoid accumulating corrosive chlorides, sulphides, etc., and also to help prevent freezing. Where heaters are required to prevent freezing, they should be checked regularly for proper operation.

As discussed in Chapter 5, air (oxygen) frequently finds its way into light ends streams and into storage vessels. The same procedures used on process units for sampling and venting vapour spaces apply to 'outside the battery limits' storage. There is one very important point to remember concerning butane, isopentane and pentane. At temperatures below 31°F (−0.5°C) for butane, 82°F (28°C) for isopentane and 97°F (36°C) for pentane, the vapour pressure of these hydrocarbons is less than atmospheric. A pentane or isopentane spheroid or butane sphere, thus, will be under vacuum. Some method of admitting an inert gas, such as nitrogen, into the vapour space is used to keep a positive pressure in the sphere or spheroid and to make the vapour space nonflammable where low temperatures may exist for long periods of time (another method could be to add propane if contamination of the product is not an issue). Air must not be allowed to enter. Remember that even though an amount of air entering at any one time may not be enough to develop an explosive mixture at the time, subsequent filling of the vessel may cause the vapour space to become explosive. The manner in which this can happen was described in Section 5.3 under *Air Hazards*.

Operators should not forget that when cylindrical horizontal vessels, spheres and spheroids are being filled or emptied, each foot on the level gauge does not represent the same number of barrels stored (Figure 35). Special charts must be used to determine how much is in storage and how many barrels may be added to reach a given level.

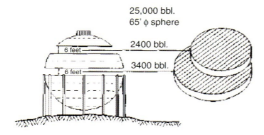

Figure 35 Every foot on the level gauge does not always contain the same volume of liquid.

ACCIDENT A serious fire resulted at a refinery when a butane sphere was overfilled. Two butane spheres ruptured in the fire, and two crude tanks were lost in an adjoining refinery (Figure 36).

Figure 36 The sphere at the left (which contained butane) was pumped over. The spilled butane ignited, and the resulting fire ruptured the sphere shell.

When filling a storage vessel, always remember to leave sufficient vapour space for the liquid to expand if the temperature increases. If a vessel is overfilled, the relief valve may 'pop' when the temperature rises (Figure 37). Remember that a small amount of vented liquid vaporizes into a large amount of vapour (see Figure 2 on page 2).

Figure 37 If a vessel is overfilled, the relief valve may 'pop' when the temperature rises.

Refer to Appendix 1 for regular checks to be conducted during routine operation of LPG facilities.

7

Handling LPG at truck transport and railroad car terminals

Since LPG must be stored and moved under pressure, proper procedures must be followed and proper equipment used. However, by strictly following safe practices, it can be loaded, transported and unloaded safely in railroad tank cars (Figure 38) and tank trucks (Figure 39).

Figure 38 LPG rail car.

Figure 39 LPG truck transport.

Most refineries do not control tank cars or trucks in transit between terminal points. However, many refineries do load and unload LPG products and must be able to do so safely. Let us discuss how this can be accomplished.

7.1 Operations common to loading and unloading tank cars and tank trucks

Several important fundamentals for safe loading and unloading of LPG products are common to both types of carriers. To handle LPG during loading operations, loaders should be aware of the following.

Loading

Efforts should be made to assure that no truck or tank car containing air is ever loaded with LPG. The following will convey some idea of the potential consequences of the practice of loading LPG trucks and tank cars which previously have been unloaded by pressuring with air.

> **ACCIDENT** While a truck was being loaded with LPG in a refinery, gas displaced from the truck was vented to a low-pressure gas system. The gas line became red hot at the point where the vent entered the line. Subsequent investigation disclosed that the truck had last been emptied with compressed air, and that the gas line contained pyrophoric iron sulphide. During the loading, the iron sulphide ignited the air/LPG vapour mixture which was being vented into the line.

Compressed air should not be used for unloading. However before reloading, if in doubt that a truck has been unloaded with air, the hazard can be reduced if a sample of vapour (Figure 40) from every truck or tank car to be loaded with LPG is first checked for oxygen content with a portable oxygen analyzer (Figure 41).

Figure 40 Vapour sample being taken from an empty LPG truck.

Figure 41 Vapour sample from empty LPG truck being tested for oxygen content on a portable-type oxygen analyzer.

No LPG should be loaded until oxygen has been reduced to an acceptable level. Specific instructions are provided for this work.

Containers must be designed for the particular type of LPG product to be shipped. Propane (because it has a much higher vapour pressure than butane) should never be put in a transport vehicle designed strictly for butane (Figure 42).

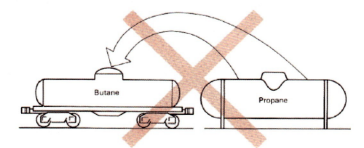

Figure 42 Never put propane in a butane container.

Loading should not be started before a check has been made to determine if any liquid remains in the carrier. If liquid is found, the car should not be loaded until the liquid is identified by a laboratory analysis or operating management has otherwise made certain its identity.

Never fill an LPG tank car or truck completely full. If the temperature of the contents rises in a full container during transit, the relief valve will vent liquid due to the pressure built up by liquid expansion. Very little liquid butane has to be vented to create a large cloud of vapour (see Figure 2 on page 2).

Outages left in tank cars and trucks should conform to the regulations.

Check the level frequently while loading. The amount of LPG in a carrier can be verified by the weight check normally performed before shipping.

Use hoses that are designed for the LPG products to be loaded. If in doubt, loaders should be sure to check with their supervisor. Hoses should be hydrostatically pressure-tested at least once a year, and in the interim period, both hoses and fittings should be inspected regularly for cuts, cracks, rusted clamps and other signs of damage.

Care must be taken so that carriers are not overpressured during filling.

As liquid flows into a container, it compresses and condenses the vapour in the container in order to make room for the incoming liquid. Both the compressing and condensing produce heat which increases the pressure in the receiving container. One method of preventing this pressure build-up is to vent the receiving container back to the container being emptied. If this method of relieving pressure is used, it is preferable not to open the vent until the pressure in the container being filled is more than that in the container being emptied. No matter what method is used, however, the loader must keep a close watch of the pressure gauge on the receiving container to prevent overpressuring.

Be sure that odorant is added to the product if it is required.

It is important to regularly check that the threads of hose couplings are still within their original specification. Excessive wear has been the cause of many fatal incidents (see the following example). Only certified adapter pieces should be used. These provide additional thread connections which will require inspection and certification. It is suggested that only refinery/terminal tested and certified adapters are allowed into service.

ACCIDENT A LPG truck driver suffered severe burns from a fire at a refinery LPG loading facility. After two thirds of the LPG truck were loaded the loading arm detached from the tank truck nozzle and propane was released and ignited (Figure 43). The loading procedure was supervised by an operator who immediately attended to the burning driver, pressed the emergency shutdown button and extinguished the fire in the loading area. The truck driver suffered severe 2nd and 3rd degree burns and later died. The quick-action valves of the loading arm closed by gas detector alarm after 4 seconds (released liquid propane volume max 55 l (14.5 US gallons)). The cause of the ignition remains unclear, although there were three possible sources:

- mechanical spark from the wrench tool;
- mechanical spark from the loading arm hitting a metal casing during the backswing movement (note that enclosed areas such as the one shown below should be avoided);
- electrostatic ignition.

Figure 43a LPG loading station after fire.

Loading arm coupling

Truck coupling

continued

The connection between the loading arm and the truck detached because of a weakened connection, despite a previous standard leak test carried out by the driver and loading operator. Both the thread of the loading arm's coupling nut and the thread of the truck connection showed significant deviations from the manufacturing standard including:

- inconspicuous contour change of the trapeze-shaped convolutions (worn down flanks);
- inconspicuous wear of 1 to 2 mm (0.04 to 0.08 in) each in diameter, therefore only minor thread overlap.

Lessons learned
- No use of hammers for fixing thread type LPG connections—this will cause and accelerate deformation of threads.
- Provide spark-free and fit-for-purpose wrench tools for LPG thread type connections and remove unsuitable and non-spark-free tools from service. Non-sparking tools need to be regularly checked to make sure that they have not become degraded by bits of grit or iron embedded in the working surfaces.
- Use thread test kit (or equivalent tools) to continuously monitor thread measures on both refinery and truck sides (see Figure 43b), identify wear and tear and remove unsuitable connectors from service immediately.

Figure 43b Thread test kit.

Unloading

Before unloading, loaders should check the shipping notice to be certain that the correct quantity and type of product are contained in the carrier.

During cold weather, the temperature in a tank car or truck transport may drop below 31°F (0°C). Below this temperature, butane's vapour pressure will be less than atmospheric, and unloading may be difficult due to the vacuum present. Air should not be used to pressure the container! Use refinery inert gas if it is available at the required pressure. If it is not, use bottled nitrogen, being careful not to overpressure the tank. Another desirable practice is to route some of the butane from the unloading-pump discharge through a heat exchanger and back into the vapour space on the tank car or truck.

Propane vapour can also be used, if the propane which dissolves in the butane does not harm the subsequent use of the butane.

Figure 44 Example of road carloading station using flexible hoses.

Loading and unloading

Before an LPG hose is completely disconnected after use, gas should be vented from the hose to reduce the pressure in it and to prevent it from swinging about violently. It is desirable to vent the hose to atmosphere through an elevated vent stack. Make sure that all hoses are properly disconnected when the unloading operation is complete and before moving the truck transport or tank car. Figure 5 on page 3 shows a serious fire that resulted when a truck driver forgot to disconnect the fill hose.

Relief valves are mandatory on all LPG truck transports and tank cars in most countries. A loader should never try to adjust or tamper with a safety relief valve. Vents from relief valves should be directed up and away from the truck transport or tank car.

Example of a firefighter training simulation of a pool fire warming a LPG truck vessel and the relief valve lifting (note that the released product from relief valve has not yet ignited).

ACCIDENT A serious accident occurred in Oklahoma City, Oklahoma, due to improper safety relief valve venting on an LPG tank truck. A fire of undetermined origin occurred at the rear compartment of the trailer. The fire heated the tanks, causing the discharge of fuel from the propane relief valve. The safety relief valve vent discharged into the enclosed rear compartment, and an explosion occurred. Five people were killed and 21 injured (Figure 45). The resulting fire destroyed a lumber yard, house, shed and an automobile.

Figure 45 *An improperly installed relief valve vent on this truck was the apparent cause of a serious explosion and resulting fire.*

Regulations covering the transport of hazardous materials have been defined and published under Title 49, Code of Federal Regulations in the US; the European Agreement concerning the International Carriage of Dangerous Goods by Road 20032 (ADR 2003) and Regulations concerning the International Carriage of Dangerous Goods by Rail 20033 (RID 2003). There are similar rules in other countries. These regulations have real meaning and purpose to protect the public and penalties for violation can be severe.

7.2 Operations concerning truck transports

No transport should be allowed to approach the rack while another transport is loading or unloading.

Loading-rack or other knowledgeable personnel should inspect each transport for leaks and other obvious mechanical defects. Leaks of any kind should be reported to the supervisor. Loaders should also be sure that the proper equipment is on the tanks and that it is in good working order.

Some truck transports are equipped with pumps for unloading that are driven directly by the truck engine. The use of such pumps for unloading at the refinery LPG rack should be avoided whenever possible. An explosion-proof electric motor-driven pump, provided at the unloading facilities, is preferred.

Approval to use power take-off pumps should be obtained from management after a careful study of the accident exposure. Consideration must be given to such items as area congestion, personnel exposure, ability to limit any difficulty to the unloading facilities, and available fire protection equipment.

Be sure that the transport tanks are level before loading or unloading. If the tanks are not level, the relief valve connection may be flooded (Figure 46).

Should the tank be overpressured for any reason under these circumstances, liquid would be vented. It is also important to keep the tanks level to obtain an accurate gauge of the contents.

LPG transport trucks are subject to maximum filling density regulations.

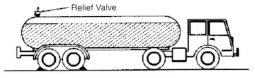

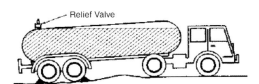

Figure 46 Make sure trucks are level at loading racks. Avoid flooded relief valves.

The maximum is based on the loading temperature and the specific gravity of the product being loaded, and can be found in a group of charts or tables usually located at the loading area.

Three different methods are in use to measure the amount loaded into a truck transport:

- rotary gauge on the side of the tanker;
- scale weights; and
- temperature-corrected meters.

Any one of the three may be used.

The following are some of the important points for loading the tankers. When the truck is in position, all electrical systems should be shut off to eliminate ignition sources. The truck wheels should be chocked and the tanker earthed. The liquid and vapour hoses should be connected, with careful observation for leaks. A knowledgeable person must stand by during the entire loading or unloading operation. Odorant must be added if required, and proper documents and placards must be provided. Assurance that the loading hoses are disconnected must be obtained before the truck leaves the rack.

A sign in large visible lettering at the loading area can be helpful. Such a sign would summarize the loading and unloading procedure, with additional detail in a procedures manual at the loading rack.

7.3 Operations concerning railroad tank cars

Figure 47 Example of rail car unloading station using flexible hoses.

Loading and unloading procedures must comply with regulations. In addition to safety problems within the refinery which must be resolved satisfactorily, there are other safety problems outside the refinery for which the refinery bears great responsibility. The regulations are designed to assist in this. Special instructions to the loader-unloader will assure that both of these needs are satisfied. It is to be expected that inspectors from the local authorities (such as the Department of Transportation in the US, the DRIRE in France or the HSE in the UK) will inspect the facilities and review the loading procedures that must comply with local regulations (such as the ADR-RID in Europe).

The following are some of the important points in loading tank cars:

- The pre-trip inspection must be done and any defects remedied.
- The tank test date and the relief valve test date must be checked to ensure these dates have not been exceeded.
- The suitability of the car for the product to be loaded must be verified.
- The handbrake must be set and the wheel chocks applied to prevent car movement.
- The stop sign must be in the proper place on the track.
- When these steps are completed, the liquid and vapour lines can be hooked up and the gauge rod set for about 30 inches (76 cm).
- When product loading is started, the sample line should be opened momentarily to remove water from the car. The gauge rod (when installed) should be fitted with the liquid orifice.
- The liquid temperature of product entering the car should be measured using the tank car thermowell and a suitable thermometer.
- The permissible loading level must be calculated, after which the gauge rod (when installed) can be reset to the outage of the particular car being loaded.
- Odorant should be added as needed.
- When loading is complete, all valves should be closed.
- After 15 minutes, the gauge rod should be raised one inch or more, the valve checked for vapour, and the gauge rod then lowered until white mist appears. The gauge rod should be read and the gauging operation repeated. The gauge should be recorded to the nearest 1/4 inch.
- The gauge rod and thermowell (when installed) should be secured.
- The hose connections should be removed and connections plugged.
- The decal should be in place inside the dome cover, giving the 'After Unloading' instructions.
- The dome cover should be closed and sealed, and the Department of Transportation placards placed in the four holders on sides and ends.

For unloading tank cars, if gallonage is needed for unloading purposes, measure temperature of contents, raise gauge rod to full extension, and depress slowly until white mist is expelled. Use temperature reading, gauge-rod reading, specific gravity of the product and the outage tables of the tank car to obtain the gallonage. After unloading, all connections should be secured, and the instructions on the decal 'After Unloading' (inside dome cover) should be followed.

In many cases, the tank car lessor (Union Tank Car, for instance) provides a booklet on suggested procedures for loading and unloading tank cars. This is a valuable reference.

8
Design considerations

8.1 Materials

Cast iron should not be used for equipment parts containing light ends under pressure. This includes pumps, vessels, piping and fittings. Commonly overlooked are equipment closure items, such as packing glands, mixer seal housings and compressor valve covers. Preferred material in these cases is steel. Cast iron is very brittle and is easily broken if it is hit sharply or cooled rapidly with water during a fire.

At one refinery, the quick closing of a valve on the discharge side of a cast-iron pump caused a pressure surge which ruptured the pump (Figure 48). Cast steel pumps are much more suitable for withstanding such pressure pulsations.

ACCIDENT

Figure 48 Cast iron pump casing ruptured by pressure pulsations.

ACCIDENT In 1948 in Texas City, two breaks occurred in the 6,800 foot (2 km) long line used for transferring a propane-propylene mixture (under pressure of 250 pounds per square inch). A low hanging cloud of gas covering 68,000 square feet (6,317 m^2) formed and drifted over the Texas City-to-Galveston Highway before it flashed, killing seven persons and injuring fourteen others (Figure 49).

continued

Figure 49 Breaks at cast iron finings in a propane-propylene pipeline caused this Texas City-to-Galveston Highway flash fire.

One break occurred at a cast iron tee and the other 748 feet (70 m) away in a cast-iron pipe collar. Both leaks are believed to have been caused by excessive stresses in the pipeline. *Welded steel piping should be used in light ends service*. Large screwed pipe couplings and valves fail very quickly when exposed to fire (Figure 50). Such failures probably will add more fuel to the fire.

Figure 50 Screwed pipe failure due to fire.

A small amount of corrosion will start leakage through the threads of a plugged nipple, especially when light ends are being handled. A suitable alloy should be used for screwed plugs used in drains, vents, carbon-steel lines, pumps, exchangers and other equipment in which light ends are contained.

Materials (such as brass, bronze and aluminum) which will not withstand the high temperatures reached during a fire should be avoided for equipment handling light ends.

Small bore screw fittings on pressure vessels and lines should be eliminated whenever possible. No small bore screw fittings should be permitted to be directly connected to a pressure source. The concern is that the threads will strip under pressure or be incorrectly screwed on during maintenance work, resulting in failure. Small bore connections (including sight glasses) should be fitted with flow restrictors to limit the rate of release in case of failure.

Flanges should be minimized where possible by welded piping to reduce leak potential.

Material characteristics should be able to withstand all service conditions, including start-up and shutdown but also colder conditions created by escaping LPG in case of a leak. Materials in contact with LPG can experience relatively low temperatures, even at normal pressures and ambient conditions. Rapid offtake of vapour from a vessel can result in low temperatures within the vessel. Similar low temperatures can be produced in 'pressure reducing' equipment and it is essential that both metallic and non metallic components in the system are made from materials that retain their basic properties over a sufficiently wide temperature range.

ACCIDENT The loss of lean oil flow in a gas treatment plant created a major reduction in temperature (temperatures in parts of the plant fell to −48°C/−54°F) of an LPG heat exchanger, causing embrittlement of the steel shell. That heat exchanger ruptured, releasing a cloud of gas and oil. It is estimated that the cloud travelled 170 metres (560 ft) before reaching fired heaters where ignition occurred. After flashing back to the point of release, flames impinged on piping, which started to fail within minutes. A large fireball (Figure 51) was created when a major pressure vessel failed one hour after the fire had started. It took two days to isolate all hydrocarbon streams and finally extinguish the fire.

Figure 51a Fireball during the LPG plant fire.

Figure 51b Ruptured heat exchanger.

Two employees were killed and eight others injured. The incident caused the destruction of one plant and shutdown of two others at the site. Gas supplies were reduced to 5% of normal in the area, resulting in 250,000 workers being sent home across the state as factories and businesses were forced to shut down.

8.2 Equipment

Direct-fired heaters should be avoided whenever possible on light ends units.

When it is necessary to use a direct-fired heater, it should be located far enough away from the nearest equipment containing light ends so that there is very little possibility of flammable mixtures reaching the heater.

Electrical equipment. A full hazardous area classification should be carried out and only equipment suitable for use in the zones defined by the assessment installed. The BP Process Safety Booklet *Hazards of Electricity and Static Electricity* covers in detail the proper safeguards for electrical equipment and area classifications.

Eliminate enclosed areas. Pumps and compressors should not be placed in a totally enclosed area. If they are, even a small leak can gradually build up the concentration of hydrocarbon vapours until the room contains a flammable mixture. Exhaust systems can remove hydrocarbon vapour effectively in such installations, but it is much better to locate the equipment outside where the danger of accumulation is virtually eliminated.

Flare stacks should be of sufficient height and be located far enough from equipment containing light ends that the flare will not provide a source of ignition. The stack height also should be such that in case of a flameout, gas issuing from the stack (if heavier than air) will not reach the ground in sufficient concentration to be flammable. The base of flare supports should be fireproofed.

Knockout drums designed to drop out liquids before they reach the flare, should be provided.

There are ***instruments*** on the market which can detect the presence of an explosive mixture and will sound a warning whenever one develops. These devices can be used to warn vehicles not to enter an area where gas is present. They also can be used at locations on units where leaks might otherwise go undetected for a period of time or where railroad tracks and roads come close to the unit battery limits. Detectors need regular calibration, can be poisoned, and failure modes may be different from one type of detector to another.

No instruments leads containing hydrocarbons should be brought into control rooms.

Sight glasses. New installations should use magnetic followers (or similar technology) instead of sight glasses.

LPG pumps should be located well away from storage vessels and outside any vessel bunded area.

Pump seals are a potential area for loss of containment. It should be confirmed that the degree of seal release protection is appropriate to the location and that seal performance has, historically, been satisfactory.

Mechanical seals are more commonly used than packing on centrifugal pumps in light ends service because there is very little or no leakage from a mechanical seal in good operating condition. For this reason, packing should not be used. Special provisions must be made, however, to prevent excessive spillage when a seal fails. One of the more common methods used to do this is to install in the pump case a throttling bushing which has a close clearance between the bushing and the shaft (Figure 52). In selecting the proper seal for a pump, consideration must be given to operating temperature and pressure, and to the characteristics of the stock. Mechanical seals must be installed correctly. Most trouble with seals is caused by improper installation or poor operation.

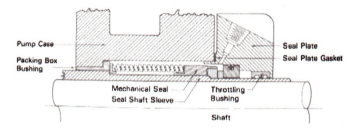

Figure 52 Typical pump mechanical seal details. Throttling bushing reduces leakage if the mechanical seal fails.

Good industry practice rates the order of protection for pump seals as shown below:

LOWEST

- Single mechanical seal;
- Single mechanical seal with auxiliary sealing device (throttle bush, lip seal or packed gland etc);
- Other configuration (2 seals);
- Double mechanical seals. These must have a seal fluid at a pressure higher than the process/system pressure.

HIGHEST

Other configuration seals are normally tandem seal arrangements where the seal fluid is at a pressure less than that of the process/system pressure or where there is no seal fluid in the case of dry running outer seals.

Caution must be exercised when terminology for seals is discussed. For further information, refer to API 619.

The need for permanently installed condition monitoring instrumentation should be assessed particularly where pumps are sited in not normally manned or unmanned locations. Otherwise, a regular program of condition monitoring is needed.

Sewers, which may receive light ends, should be suitably trapped at all inlets. Sewer boxes should be vented and snuffing steam provided for vents out of reach of a conventional steam hose-steam lance. Control-room floor drains must not be connected to the plant sewer system.

Sufficient space around a light ends process unit should be allowed to minimize the possibility of vapours from a spill drifting to a source of ignition at another unit. Roads around the units should be elevated, where possible, to prevent this from occurring. Such roads serve as firewalls and, to a degree, as vapour and liquid barriers.

Water-cooled exchangers present a problem in light ends process units. Usually, cooling water is at a lower pressure than is the process stream in the exchangers, so if a leak develops, the hydrocarbon leaks into the water. This means that light hydrocarbon, usually vaporized, will go to the sewer, another unit or a cooling tower. Serious fire hazards can thus be created at these locations. At one refinery, a cooling tower burned down because light hydrocarbon had leaked into the cooling-water system and was ignited at the cooling tower.

On one alkylation unit, the process cooling water goes to a separating drum equipped with a vent stack before going to the sewer (Figure 53). The vent stack is high enough to ensure that any reasonable amount of hydrocarbon vented will be diluted by air to the point where it will not be flammable before reaching a source of ignition. Leaking exchangers are discovered by observing the vent stack at regular intervals for signs of hydrocarbon vapour.

Figure 53 Gas separator drum for venting light ends from cooling water.

Sampling

In order to obtain a representative sample and avoid excessive purging, sample points should preferably be located on circulating systems.

The most common type of sampling system used by oil companies is shown in Figure 54 below.

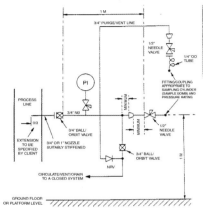

Figure 54 Sampling system.

Older common types of sampling points are of a simpler design. Figure 55 shows this simplistic design feature.

These should be assessed to see if they are adequate to minimize the possibility for loss of containment. The method of LPG sampling and of sampling storage in the laboratory should also be checked.

For the sample point shown in Figure 55, the following checks are recommended:

- check if the sampling cylinder is fitted with an ullage tube;
- check if the sampling cylinder is registered as a pressure vessel;
- check if the sampling procedure is readily available, is clear and understandable;
- check if the sampling point is labelled and that samplers are adequately trained in the correct procedures.

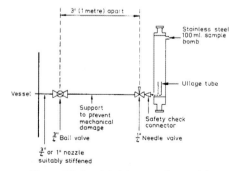

Figure 55 An old style sampling point.

Remote isolation valves. Manual only valves may be inaccessible under fire conditions and therefore remote operated vessel valves should be provided. Emergency isolation valves should be fitted on all pipework (except piping for relief valves and level measuring devices below 3 mm (0.125") for liquid and 8 mm (0.333") for vapour) connected to LPG storage vessels containing more than 10 m^3. Smaller vessels should be considered on a case by case basis.

The location of the remote operating controls should be outside of any fire area or high radiant heat area caused by fire. These controls should also be located upwind of the vessels in case of a gas leak. Additionally, the facility to close the valves automatically (such as a fusible link) in the event of a fire should be installed.

The valves should be fire-proofed and located as close as possible to the inventory (if possible inside vessel, if not welded on vessel outlet pipe, or on 1st flange after vessel). When possible, it is good practice to locate the first flange on liquid lines far away from the storage vessel, so any fire from a flange leak will not impinge on the vessel (see picture below).

First flange

Note that there is no gasket, bolted flange or manual and control valve under the vessel, therefore minimizing incidents in direct proximity of large inventory. The number of liquid line piping connections to the bottom of the tank is also minimized compared to the majority of spherical tanks (compare with other pictures in this booklet) that have at least five of these which are one inch or greater, typically:

- 6-inch fill line;
- 10-inch discharge;
- 4-inch recirculation;
- 2-inch water draw-off;
- 1-inch sampling line.

Also note passive fire protection of legs and bottom line, plus deluge system.

Spring-loaded valves should be provided on water drain lines. This will close automatically unless the operator keeps his hand on the handle. Such valves, sometimes called 'dead man valves', provide assurance that drain valves will not be left open if something happens to an operator or if circumstances prevent him from closing the drain.

Consideration should be given to installing shutoff valves in all liquid lines to and from each vessel, springloaded to close (held open by a fusible link). This will keep the vessel contents from depressuring through a ruptured field line if circumstances otherwise prevent personnel from blocking in the vessel.

Consideration should be given to providing a fire-hose connection in the fill lines to all pressure storage vessels. This will allow quickly establishing water bottoms in a vessel and/or displacing hydrocarbons with water in a field line.

Dewatering facility

Water drainage from vessels in light ends service can be complicated by the refrigeration effect of stocks that vaporize at atmospheric pressure. This effect was discussed and illustrated in Figure 9 (see page 10). In most cases, steam tracing or other means of applying heat to drain lines and valves will prevent freezing. Spring-loaded, self-closing valves, sometimes called 'dead-man valves' may be used on water drain lines. As these valves must be held open by the operator to drain, they give reasonable assurance that the drains will not be left open by freezing.

Direct draining of water to atmosphere should be avoided where practicable on all systems that have to be regularly operated and in particular where a large inventory of LPG could be accidentally released.

Where it is not practicable to drain water to a closed system, there should be provision for the safe withdrawal of water preferably using dewatering pots or an arrangement offering the equivalent degree of protection and fitted with a remote operated isolation valve.

In certain situations where water draining is very infrequent, it may be difficult to justify the installation of a draining system using dewatering pots and interlocking valves. However, it is the preferred system for dewatering.

During any draining operations, the dewatering pot must be positively isolated from the main LPG storage vessel in order to minimize the LPG quantity that could be released accidentally to atmosphere.

To guarantee isolation from the main vessel before draining procedure commences, the first drain valve after the pot should be interlocked with the isolation valve on the storage vessel. Figure 56 shows typical good practice design for dewatering facility for an LPG Pressure Vessel.

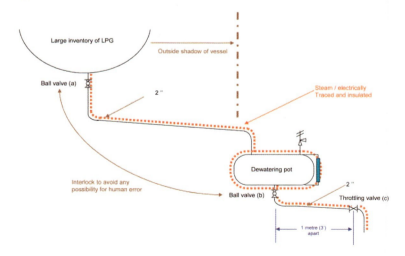

Figure 56 Dewatering facility for an LPG Pressure Vessel.

The dewatering pot and its associated piping should be protected against freezing and hydrate formation and be fitted with a relief valve for the fire case.

An all weather standing instruction for drawing off water should be written on a notice board adjacent to the draining point.

The instruction should reflect the following:

1. Valves (A), (B) or (C) will normally be kept closed.
2. Open valve (A) to drain water into the dewatering pot.
3. Close valve (A).
4. Open fully the quick shut-off valve/cock (B) immediately adjacent to the dewatering pot to be drained. In the case of an interlock arrangement, valve (B) cannot be opened unless valve (A) is completely closed.
5. Crack open the throttling valve (gate or globe) (C) and carefully drain off any water. It is possible that this valve could freeze in the open position during draining operations.
6. Close the throttling valve (C) when the draining is completed.
7. Close the quick shut-off valve (B).
8. Repeat 2 to drain more water into the dewatering pot.
9. On completion of water draining, open and close the throttling valve (C) to release liquid held up in the section of pipe between the two valves.

Where the above system is not provided, a simple two-valve system may be acceptable if a full risk assessment has been carried out. This would include a ball valve adjacent to the vessel with a well spaced ($>$1 metre/3.3 ft) throttling valve in a horizontal drain line taking water away from the vessel.

This type of system would typically be used where dewatering is carried out very infrequently. Therefore, under normal circumstances, the first ball valve should be locked closed and the end of the drain line blanked off.

It is essential that clear instructions be given prior to this non-routine draining operation which should be carried out under work permit.

The operator/technician must appreciate the significance of the correct sequence of valve operation in view of the possibility of ice plugging.

Again, following completion of draining, the liquid in the section between the two valves must be released.

Drain lines and valves for draining water from storage vessels must be protected from freezing; otherwise, water in these valves and lines may freeze

ACCIDENT A fire that could have had serious consequences occurred at a butane sphere equipped with external bottom drains (Figures 57). During a sudden and unexpected cold wave, a makeshift heating device was used to prevent the external line and drain cock from freezing. When water was drained from the sphere, some butane escaped and was ignited by the heating device. Fortunately, the drain was closed quickly, thereby preventing enough butane from escaping to cause a serious fire.

Figure 57 Unprotected sphere connections may freeze.

An internal drain connection (Figure 58) provides a solution to this very serious problem. Water can be drained away from the cock, and no external means of freeze protection is needed. Each time water is drained through this piping arrangement, it is preceded by a slug of the sphere stock. If this design is used, operators must be warned to be sure that no source of ignition is in the area.

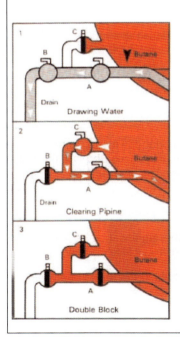

Figure 58 To operate internal water drain open valves 'A' and 'B' (1); liquid head pushes out water. When water draw is complete, close valve 'B' and open valve 'C' to permit clearing lines of remaining water (2); when lines are clear of water, close valves 'A' and 'C' (3).

and rupture them. In cold climates where the possibility of freezing exists, the drain line should be lagged (insulated) and traced up to the throttling valve. Also, water freezing in a valve may prevent its proper operation and may make it impossible to drain equipment thoroughly.

Pressure relief valves (PRV)

Vessels should be equipped with two 100% capacity pressure relief valves, designed and installed in accordance with API 2510/520/521 (preferred in oil industry, but European standards for unfired pressure vessels or BSI standard BS 5500 may be equally acceptable) and in compliance with the pressure vessel's design code.

Isolation valves fitted beneath the relief valves should be of the full flow bore type. High integrity arrangements such as an interlock system or locks and chains should be installed to ensure that at least one isolation valve is always locked in the open position with its associated 100% capacity relief valve operational.

A register of relief valves should be maintained and each individual valve identified by a works identification number.

The relief valve tail pipe should be adequately supported and covered with loose fitting plastic caps to prevent ingress of water. (The provision of vent pipe covers for the PRV's prevents rain entering and therefore corrosion build up, which may lead to corrosion particles affecting the valve. Covers also prevent birds from nesting in the tailpipe, causing potential blockages over time.)

The tail pipe should have a drain pipe provided which directs away from the vessel to prevent any flame impingement on to the shell.

Relief valves will not protect vessels against fire exposure if the temperature of the unwetted internal shell metal gets so high as to result in failure below the relief valve set pressure. Such failures result in BLEVE events. Pressure vessels are therefore typically protected against fire exposure by the following:

- slope to drain away from under the vessel for pool fire impact reduction;
- good nozzle/pipework design to minimize the possibility of jet fire impingement on the tank shell;
- water deluge systems for cooling, or;
- passive fire protection.

Figure 59 shows the relief valve requirements.

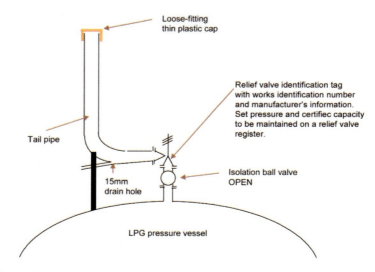

Figure 59 Pressure relief arrangement for a pressurized LPG vessel (showing only one of the two 100% capacity pressure relief valves—the other relief valve would be mounted on a separate vessel nozzle. A high integrity arrangement such as an inter-lock system or locks and chains should be installed to ensure at least one isolation valve is always in the open position).

Relief valve vent pipes should be arranged so that the flames will not impinge on the vessel shell should the vented vapours catch fire.

ACCIDENT In 1956 at a Texas refinery, 19 firefighters died when vapours from about 500,000 gallons (1,900 m³) of a mixture of pentane and hexane escaped as a huge ball of fire from a ruptured spheroid tank (Figure 60). Flames from burning vapours issued from a vent impinged on the spheroid, weakened the metal on the top plate above the vapour space, and caused the rupture.

Figure 60 The spheroid at 'A' ruptured and caused the tank fire shown. Flame impingement from a relief valve weakened the spheroid shell. The spheroid at 'B' is the companion to the one at 'A'.

Piping and relief design should also consider specific properties of some light ends' (such as butadiene) propensity to decompose in 'popcorn' polymers. Popcorn polymers are crosslinked polybutadienes. It is known that butadiene reacts with air (oxygen), rust (iron oxide particles) or water to form polymeric peroxides which are sensitive to heat and shock and decompose easily. Decomposition of butadiene polyperoxides form free radicals which are likely initiators of polymerization. The 1,3 butadiene monomers polymerize at active free radical ends and create crosslinkings. This reaction is extremely exothermic and could provide enough heat to expand and overpressure the pipe, resulting in pipe rupture.

$$nC=C-C=C \xrightarrow[\text{Promoter: } O_2, \text{ rust, water}]{\text{Polymerization and crosslinking}} (C=C-C=C)n-C-C-C(C-C=C-C)m$$

1,3 butadiene monomers >75%

$$|\\ C=C$$

[Popcorn Polymer]

ACCIDENT A butadiene unit was quickly shut down when a butadiene vapour cloud was released from a 1 m (39 inches) split rupture on the overhead pipe from the reboiler on the final purification tower. A large quantity of popcorn polymer was noticed in the area local to the leak substantiating the fact that the split was caused by the tremendous forces created during its formation. Fortunately, the released hydrocarbon vapour cloud did not ignite and no fatalities or injuries were recorded. Approximately five tonnes of butadiene-rich hydrocarbon vapour were released. The location where the pipe was split was on a safety valve line which is a dead-leg line and 'live' to the process. This line was not sloped as per pipe specifications, possibly allowing liquid butadiene to pool in the pipe prior to the incident and then decompose.

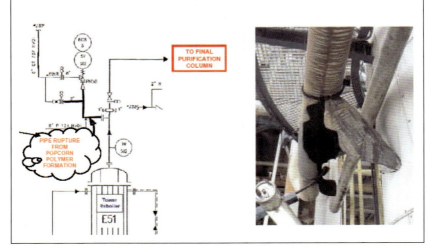

Venting capacity should be sufficient to prevent a vessel from being overpressured during a fire. Storage vessels should be designed to withstand whatever vacuum may be encountered for the stock being stored.

The danger of cylindrical horizontal storage vessels becoming rockets during a fire is described in Chapter 9. The original layouts of such vessels should be planned so that the vessels do not point toward areas where great damage may be done if a vessel fails at a circumferential seam during a fire.

Suction lines for butane and propane storage vessels should be made as short as possible. Because of the high volatility of these stocks, a small pressure drop or warming of the suction line will vaporize sufficient liquid to bind or cause cavitation in a centrifugal pump (Figure 61).

The sun, shining on blocked-in lines full of liquid light ends, can build excessive pressure. Fires have occurred when lines or gaskets ruptured. If relief valves are not provided around the block valves, a valve should be cracked slightly into a tank to relieve pressure.

SAFE HANDLING OF LIGHT ENDS

Figure 61 The sun can provide enough heat to vaporize light ends in a pump suction line.

Protection of light ends pipes from impact

Light ends pipes must be clearly identified/colour coded. They must be physically protected from impact. Underground lines must be clearly marked.

What can happen when pipelines are unprotected. Luckily no leak occurred despite a natural gas line being impacted.

Good practice—warning girder on both sides of a pipe-rack (although the first girder looks higher than the rack this is an illusion due to the picture being taken from ground level).

61

8.3 Fire protection

Local regulations and fire protection codes specify minimum spacing for many of the light ends storage vessels. The desirability of providing as much space as possible between storage vessels and sources of ignition cannot be overemphasized. Any light ends storage tank should be located so that the prevailing winds will carry the vapour from a spill away from any source of ignition.

Fire protection provisions present one of the more difficult engineering design problems on liquid butane or propane storage projects. In the UK the LP Gas Association publish codes that provide solutions for the protection of storage. Figure 62 shows an acceptable design for protecting a sphere from failure due to an area fire.

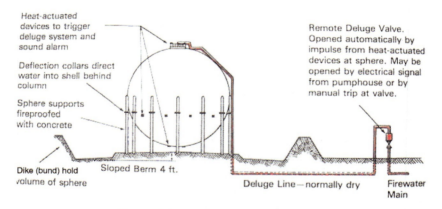

Figure 62 Typical fire protection for spheres.

Ground slope/gradient

It is necessary to create a slope under vessels to permit liquid spills to drain away from the underside of the vessel. This prevents a sustained pool fire under the vessel and therefore reduces the risk of prolonged vessel flame impingement.

The ground in the bund should be concreted or paved smooth to allow free flow. Low separation walls or kerbs (maximum 0.6 m/2 ft high) should be installed around each vessel for containment and direction of any spillage to a catchment/evaporation area via a graded ditch or drain. Stone chippings should not be used beneath vessels since they retain spillage and aid evaporation.

Fire protection at LPG storage facilities is designed on the basis that water application shall be sufficient to control the vessel's metal temperature, such that there will be no loss of strength in a fire envelopment situation (called a pool fire). The maximum heat flux in such a situation is normally assumed to be

100 kW/m², and the aim of any protection system is to limit the heat flux to 20 kW/m² in order to provide effective control of the developed pressure within the vessel.

In the event of a torching type flame impinging directly on to the metal surface (jet fire), the heat input rate can be 3, 00–700 kW/m².

Water deluge systems are intended to provide a water film of at least 0.1 mm on the metal surface, and research indicates that so long as this film is maintained, the plate temperature can be kept to 100°C (212°F) in a fire envelopment situation. LPG vessels fail at about 600°C (1,112°F), and steel starts to lose structural strength at about 300°C (572°F). To provide fire protection in a fire envelopment situation, the general recommendation is that water be applied to the vessel at a rate of 7.3 litres/m²/min (0.18 gpm/ft²); a deluge rate of 9.8 litres/m²/min (0.24 gpm/ft²) is necessary to ensure this application rate under all conditions.

The vessel is protected from ground fires by an automatic water-deluge system. These systems should be inspected and tested periodically with clear criteria for nozzle performance and flow rates to be sure that they are free of obstructions. Heat-actuated devices strategically located on the surface of the sphere may detect any sudden heat rise and, in turn, cause the remote deluge valve to open. Water from the firewater main then flows to the top of the sphere.

Water is distributed over a sphere by a weir plate (Figure 63) around the top of the sphere. A deflection collar fastened to each column, located directly above the column fireproofing, prevents water from running down the column and missing the shell plate directly behind the column (Figure 64). Sufficient water is provided to remove a large part of the heat radiated to the sphere.

For water protection to be effective, water must be provided to an empty or partially filled sphere within five minutes of a fire beginning.

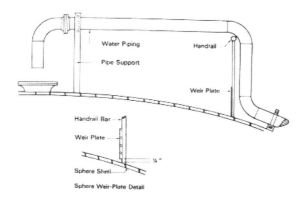

Figure 63 Details of a water-deluge system used at the top of a sphere.

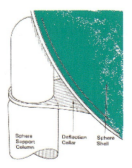

Figure 64 Deflection collar prevents water from running down the column. It directs the water onto the shell behind the collar.

This type of system has to consider manways, valves and other fittings that may prevent the water running down evenly over the complete sphere shell area, from top to bottom.

For other storage geometries such as horizontal tanks, codes such as the LP Gas Association ones can provide acceptable standards.

Another important point to note is that for cooling spray systems, it may be necessary to provide a spray ring for the lower hemisphere area, where pool flame impingement can create most impact. Some top mounted spray systems do not offer full water coverage under the lower half of the sphere and water slippage or wind effects can reduce the protection for this part of the sphere.

A deluge system should not be considered as offering protection against jet fires.

Figure 65 Example of lower fitted cooling spray rings for the lower hemisphere shell area to counter slippage and flame impingement effects.

Figure 66 Example of a fixed ring cooling spray system. The nozzles may become blocked and they require regular testing and flushing to ensure total water coverage.

Another point to note is that water spray system rings are prone to spray-nozzle blockages, mainly from corrosion particles within pipework or particles in poor firewater supplies.

Water supply should be based on a major scenario, with a minimum of two hours' supply provided.

Passive fire protection

To overcome the problem of potentially ineffective water sprays, the use of passive fire protection coatings as thermal (fire) insulation offers a good alternative for liquid butane and propane storage vessels under fire conditions.

Whilst this type of protection will reduce heat input from fire, means of applying water still should be provided to keep exposed metal (stairways, valves etc) temperatures down during a prolonged fire. If passive fire protection is to be considered, it should be recognized that corrosion of the vessel wall can occur under the insulation. Inspection procedures must recognize this problem.

Regardless of the use of passive fire protection coatings or not for the vessels, storage vessel supports should always be fireproofed with a minimum fire rating of two hours.

Note that when insulation or fire proofing is used, it should be remembered that corrosion of the vessel wall can occur under it. Inspection procedures must recognize this problem.

ACCIDENT

Figure 67 A 20-year old 12,580 bbls (2,000 m³) sphere was taken out of service for internal inspection and a hydrotest. It was approximately 75% full of water in preparation for the hydrotest when the legs collapsed. One death and one injury occurred due to the structural failure. The legs of the sphere were coated with fireproof concrete and salt water was used in the water deluge fire system on the sphere. Water sprays were tested at periodic intervals. The legs had suffered severe corrosion underneath the fireproofing. The same incident occurred at another refinery in the 1970s.

8.4 Design considerations for loading facilities

Loading facilities must be designed to be safe and efficient. The following paragraphs outline important design considerations for loading facilities.

Truck transport racks should be designed so that the possibility of a truck hitting an LPG line is minimized. Sufficient space should be provided so that a truck does not have to back into position.

Sufficient space should be allowed between loading facilities and any sources of ignition to minimize the possibility of fires or explosions if leaks or spills occur. Local regulations and fire protection codes specify minimum spacing. It is advisable to locate the loading facility so that the prevailing wind blows from any permanent sources of ignition towards the loading area. Do not forget that vehicles on public highways and locomotives on railroads can ignite LPG.

Electrical equipment that could provide a source of ignition should be avoided.

Racks must be grounded to protect against static electricity and stray currents.

Accidents sometimes occur where the flow of LPG to the rack must be stopped, yet conditions are such that the operator cannot get to the proper valves or pumps at the rack. Provisions must be made for remotely shutting off the flow of LPG. These facilities must be close enough to the rack for the operator to get to them quickly on the escape routes.

Adequate fire protection equipment should be provided. Water water-fog spray systems should be installed complemented by fixed monitors—both can also be used to dilute small leaks. Portable and trolley dry powder extinguishers should be available for small fires.

Other design features normally included to make the rack safe for operating personnel are adequate handrails and toe plates, open grating for walks and platforms, and large-sized drop platforms.

Articulated loading arms are preferable to flexible hoses for all types of loading/unloading stations (road, rail, sea).

Emergency release, or 'dry break' couplings with automatic shut-off valves should be installed on all berths handling liquefied gases for loading or unloading of barges or ships (refer to BP Process Safety Booklet *Safe Tank Farm and (Un)loading Operations* for more details on this).

Handling of LPG and light ends emergencies

If the principles for the safe handling of LPGs and light ends, as outlined in this booklet, are observed, the probability of leakage, fires and explosions will be minimized.

However, accidents do occur, and it is important to know the fundamentals of handling LPGs in emergencies.

LPG incident types

The types of emergency incidents that may occur involving LPGs are:

- un-ignited release, leading to:
- flash fire;
- vapour or liquid (or both) jet fire;
- liquid pool fire;
- vapour cloud explosion;
- BLEVE (Boiling Liquid Expanding Vapour Cloud Explosion).

The flash fire will typically last for a few seconds depending on the length/travel of the gas cloud. At these durations, a flash fire will not cause escalation through steelwork failure although they can damage cabling or cable insulation. The hazards from such fires will involve personnel being in the gas cloud area on ignition.

The jet fire, sometimes called torch fire, will occur where the release is pressurized on ignition. Jet fires can have an erosive effect as well as very high heat flux levels and can rapidly cause failure in steelwork.

A liquid pool fire may occur where sufficient quantities are released and may typically occur in a bund.

The vapour cloud explosion occurs where a large volume of gas or vapour forms in a congested area, typically in process plants or other closely confined equipment areas. Whilst conditions for an 'explosion' have to be correct in terms of air/fuel mixture in the vapour cloud, the overpressures and consequential damage can be devastating. Vapour cloud explosions may occur where the vapour cloud is bigger than 5 tonnes (or smaller in very confined spaces).

A BLEVE is commonly thought to be an event limited to LPG storage spheres and storage vessels but it can also occur in processing vessels and drums.

In its simplest description, it is associated with the rupture, under fire or high radiant heat exposure conditions, of any pressure vessel or container, containing liquefied gases or any superheated flammable liquid. The resultant fireball may last from a several seconds to more than 20 seconds depending on the tonnage of LPG within the container.

9.1 Behaviour of light ends spills and fires

Liquified Natural Gas burns with no, or almost no, smoke. Its burn rate is four times that of gasoline. As more product is burnt by unit of time, the radiant heat is higher for LNG than liquid hydrocarbon.

	Burn rate of liquid pool
LNG	12.5 mm/min
Propane	8.2 mm/min
Butane	7.9 mm/min
Gasoline	4.0 mm/min

All things equal, an LNG spill will produce two to three times more vapours than an LPG spill of the same size:

- 1 volume of liquid butane will produce 222 volumes of vapours;
- 1 volume of liquid propane will produce 290 volumes of vapours;
- 1 volumes of LNG will produce 620 volumes of vapours.

LNG spill

A LNG spill on the ground will first produce 3 m^3 of vapours per minute and per square metre of spill area. Once the ground has frozen and is not able to transmit calories to the LNG, the vaporization rate slows down to 0.3 m^3/min/m^2. Therefore, a confined spill reaches a steady-state of vaporization quite quickly and is easy to control.

A spill of LNG on water will first produce 213 m^3 of vapours per minute and per square metre of spill area. Once the water has frozen and is not able to transmit calories to the LNG, the vaporization rate slows down to 0.3 m^3/min/m^2. However, a spill on open water will continue to bring calories to the LNG. It is therefore vital to drain water from bunds and other areas where LNG can spill.

For more details, refer to 'Liquefied Gas Fire Hazard Management' by SIGTTO, June 2004 ISBN 1 85609 265 7.

9.2 LPG incident response strategies

Un-ignited release

It should never be assumed that wind direction and speed will remain constant during gas or vapour release incidents and it is therefore always prudent to assume that ignition will occur.

After proper initial and ongoing risk assessment for hazards to personnel, consideration may be given to setting up water curtains or water spray to dilute the cloud to its LFL or to contain in place or to disperse the gas.

Use of fixed water curtains or monitors is preferred but these are always difficult to locate without knowing the conditions at the time of a particular incident (location of leak, wind, etc.). Where a manual response is required, deployment of portable water curtains will need to be supported by hose handline teams using water curtains to protect the deployment teams. Such work needs careful pre-planning and exercising.

Figure 68 Fire crews move in to deploy water curtain nozzles under hose handline water screen protection.

The full response strategy for unignited gas or vapour release events depends on the size and nature of the release and the wind direction, but the overall strategy should always consider the following:

- halting all work;
- halting, switching-off and abandoning vehicles and machinery on site or in the affected area;

- evacuation of all non-essential personnel;
- isolation of the release source, if possible, safe and practical to do so. (Avoiding the use of any ferrous or aluminium hand tools for valve or handle closures);
- earliest assessment of gas/vapour cloud extent and areas that may be affected either downwind or on all perimeters of the cloud;
- earliest identification of potential ignition sources the gas may envelop and actions including heaters, naked lights extinguished, electrical isolation (Only if this is possible in a safe time frame before any gas migrates to the equipment);
- road traffic on nearby public roads;
- pedestrians/passers-by;
- non-use of telephones or radios in the suspected/confirmed gas hazard perimeters;
- alerting any neighbouring industrial or public areas to the gas hazard;
- road traffic closures;
- actuating any external emergency plans to evacuate people and halt traffic, etc;
- emergency response from upwind direction;
- arrangements for meeting oncoming external response groups at a safe distance from the facility to advise gas cloud affected location;
- awareness of plant/process drainage systems, storm water drainage and other low level drainage or piping, conduit, ducting or cable sleeve routes through which gas may migrate to remote locations;
- awareness of buildings and enclosures where gas may accumulate and then ignite, causing an explosion;
- use of water in the form of water screens/curtains/spray that may dilute, check, contain or minimize gas migration;
- use of portable gas monitoring equipment to monitor gas cloud extent;
- use of water streams to agitate/aid gas dispersion at or near to the release source. (This would only be practical if the release was not of significant size or scale and the streams were placed close together at the source to create turbulence);
- avoid entering the gas hazard area under any circumstances;
- expect gas ignition at any time, even if there does not appear to be any ignition source.
- awareness that if a large gas cloud does ignite in an equipment congested or plant congested area, a vapour cloud explosion may result with resultant overpressures and blast debris consequences.

ACCIDENT

Figure 69 This example of an unignited gas release involves a ship but the same principles apply. In this case, LPG was released when a loading arm accidentally disconnected. The extent of vapour migration is clearly shown. Luckily, the cloud did not ignite.

ACCIDENT

Figure 70 A major gas leak occurred on a platform when an O-ring seal failed on the door of a vertical pig trap—the gas cloud drifted over the platform without finding an ignition source.

The best method of water application for gas/vapour dilution or dispersal, where it is safe to move close to the release source, is to set monitor nozzles to a wide angle spray and gradually build up water pressure, whilst trimming the water spray to a semi-fog or semi-solid stream pattern that will create turbulence to assist in the rapid dispersal of the gas to its LEL.

Figure 71 Monitors set to a wide angle spray to assist in the dilution of a gas release near the source.

Water flooding connections

The provision of a water flood connection on the LPG product fill line connected to the bottom of LPG storage tanks should be considered, with adequate valving and non-return valve. This type of provision provides the benefit of displacing LPG with water if an accidental release of LPG liquid occurs from a tank bottom piping connection, flange, valve or fitting which cannot be readily isolated. Pressure monitoring should be clearly addressed in this contingency pre-planning to ensure that LPG cannot back flow into water piping.

Flash fire

The flash fire will occur with vapour cloud or vapour/gas release ignition and rapid burn-back to the release source if the release is continuous. This may cause a residual jet/torch fire at the release source if sufficient pressure remains.

Typical flash fire events will see ignition at the cloud 'edge' followed by physical burning as the flame front develops and travels through the cloud.

There is no response strategy for this type of incident, other than evacuation from the gas cloud area to prevent injury, in anticipation of ignition; and a possible dilution/dispersion of the cloud before ignition.

ACCIDENT A drainage valve of one of the three 100 m³ (26,400 US gallons) LPG storage vessels at an LPG depot was leaking and needed to be replaced. The product was transferred to another tank and the gas phase released to the atmosphere. The plant manager reopened the valve several times to drain the remaining liquid (which he assumed to be water) from the tank bottom. A gas cloud developed and a flash fire ignited on an electrical panel in the area of bottle sales. A customer and a member of the sales staff were severely injured and transferred to a specialized hospital.

Figure 72a Draining point on the LPG vessel.

Figure 72b Electrical panel.

Figure 72c Burnt grass at the entrance of the depot.

Jet fire

Of all the fire events that are encountered by responders, the jet fire has the potential to cause rapid escalation through total failure of a sphere or vessel or drum, with subsequent BLEVE consequences.

Figure 73 Bending a small jet fire on a liquid hydrocarbon loading rack (training at BP fire school).

Under certain conditions, it may be possible to 'bend' a minor to moderate gas jet fire or to deflect it away from a sphere or vessel. This particular tactic has had some success but only on relatively low pressure jet fires. The tactic employs several water streams that are set to a pattern between semi-fog and straight jet.

The aim is to hit a large area of the flame near the impingement area and push it away from the equipment. It is stressed that this tactic involves responders approaching close to the fire area using hose handlines. Therefore, if this is considered as a strategy, the water application must be deployed early on in the incident.

Under no circumstances should any such attempt be made where flame impingement has been ongoing for more than 10 minutes.

In every case, it is necessary to establish, as early as possible, the contents of the vessel or container, the fill level, time of fire starting and if possible, the existing contents status. This will aid risk assessment and allow decisions to be made on sound information.

The objective is to apply water on to flame or heat affected exposures until either isolation or burn out of the jet fire is achieved.

Jet fire response strategies may also need to consider several important safety factors:

- the level of awareness/training/experience of the responders, whether in-house or local fire service;
- to be effective, water has to be applied in copious quantities directly on to the area of jet flame impingement on affected LPG containers;
- responders must be donned in full fire resistant personal protective equipment;

- use of wheeled or portable firewater monitors to reduce personnel exposure hazards;
- manhandling firewater monitors into position under protection;
- delays in setting-up or applying water must consider rapid escalation potential and therefore full responder evacuation may be necessary;
- do not attempt to extinguish the jet fire by using dry powder or heavy water application;
- prolonged water application may result in drainage problems or flooding events that may undermine hardstandings, pads or sphere or vessel base areas.

If there is ever any doubt over the extent of flame impingement on containers or on the condition of these, consideration must be given to full evacuation of the area unless water can be directed to the affected plant in a timely manner.

There are no clear time warnings ahead of container failure, where flame impingement has occurred. If responders confirm flame impingement, an immediate concern must be to apply water on to the impinged or heat affected area of the plant. If this cannot be done within the first few minutes of the fire incident starting, evacuation must be given priority.

This becomes very important where no local firewater for cooling can be applied, for whatever reason, until the arrival of the local fire service after 5–10 minutes of the fire starting.

One method of assessment for cooling exposures is to play a stream of water on to suspected heat affected vessels or drums or equipment and observe the reaction. If the water vaporizes (steams), then further cooling is obviously required. If there is no steam produced, cooling is not required at that time, but further checks may be necessary.

Figure 74 Jet fire example showing the high heat and pressure flame impact on a valve. Temperatures greater than 1,300°C are typical, resulting in rapid failure of steelwork.

Liquid pool fire

For light ends in the liquid phase, including condensates/pentane, firefighting foam can be used to reduce vapour emission or to decrease radiant heat. However, the volatility of some condensates may be such that vapour can still erupt above the foam blanket. Therefore, although extinguishment can be achieved (often by a combination of foam and then dry powder), it must never be assumed that the foam blanket is achieving 100% vapour suppression.

Figure 75 Example of an LPG pool fire. High levels of radiant heat are present. Little or no smoke is produced in comparison to the 'heavier' hydrocarbon fuels.

For LPG liquid releases, foam has been used in a few instances and has been successful. The foam forms an ice sheet over the liquid LPG. The application of foam over liquid LPG spills is largely unproven and therefore this aspect of control strategy must be carefully considered. High expansion foams have the best control over LNG spills as demonstrated by recent tests.

Also, once foam application has commenced, any water streams in the immediate foaming area must be closed down to prevent foam blanket dilution.

Foam blanketing for vapour suppression

This response strategy may be used where a liquid light ends pool is forming due to a release and where the liquid release is constant and vapour migration is a serious or major hazard. The objective should be to apply foam on to the liquid to reduce vapour and thus reduce the distance to LFL.

Response strategies that may include foam blanketing need to consider several important safety factors:

- medium expansion or high expansion foam concentrate and application equipment to be used;
- upwind approach to the release source;
- handheld hoselines with branches to cover those approaching to set-up foam pourers or monitors;
- responders must be donned in full fire resistant personal protective equipment (fire turnout gear);
- manhandling foam pourers or monitors into position;

- foam application devices directed to apply foam ahead of any liquid in a rolling application to minimize static generation and ignition;
- avoid foam stream application directly into the liquid;
- safely re-accessing the foam pourers or monitors if there is a need to reposition;
- use of portable gas monitoring equipment to monitor gas extent;
- actions to be taken to deal with the spill once the foam has been successfully laid.

The applied foam blanket will need regular 'topping-up' to maintain vapour suppression. Once portable equipment is in position and functioning correctly, responders should retire to a safe distance from the general hazard area and observe the foam application from a distance.

Figure 76a Example of foam blanket over an LNG spill. This foam was applied by fixed foam pourers but, for condensate spills, portable foam application will achieve the same results if the response crews are well trained.

Figure 76b Another example of foam blanket over an LNG spill. This foam was applied by portable equipment, but in this case note the vapours migrating in the centre of the photograph.

Foam application may need to consider environmental impact and therefore this strategy should be reviewed in advance to ensure that such an impact is either eliminated or reduced. In some countries, there may be a need to contain all firewater and foam applied for incident control.

Use of water curtains or spray around a liquid release may, in some cases, be the best response to contain gas or prevent migration or aid early dispersion.

Use of waterspray/water curtain

Response strategies that may include gas dispersion from a liquid pool using water spray need to consider several important safety factors:

- upwind approach to the liquid source;
- responders donned in full fire resistant personal protective equipment;
- use wheeled or portable water monitors or water curtain nozzles (no permanent manning);
- manhandling firewater monitors/water curtain nozzles into position;

- handheld hoselines with water curtain streams to cover those approaching to set-up water curtains/screens;
- causing ignition through an electrostatic discharge from the water jet directly into the LPG liquid;
- safely re-accessing the monitors or water nozzles if there is a change of conditions (wind, release volume increase etc);
- use of portable gas monitoring equipment to monitor gas extent.

It is very important *not* to place total confidence in any waterspray dispersion of a gas release. Monitoring of the area in the vicinity of the release outside the waterspray area needs to be continuous to check if the tactic is effective.

If the water supply has to be provided by fire vehicles, then these have to be located upwind and preferably at a minimum distance of 100 m from the identified release source. Wind direction and speed monitoring are critical during such operations.

Figure 77 Example of a portable water curtain nozzle in use.

Figure 78 Fire crews laying down a water curtain nozzle. The team is fully protected against a flash fire event by hose handline water screens during the deployment.

Figure 79 An example of a fixed water curtain nozzle (above) and a water barrier in line to prevent gas/vapour migration (right).

Vapour cloud explosion

The overpressures generated by a vapour cloud explosion will depend on the quantity of gas/vapour in the cloud, the congestion of plant or equipment and the efficiency of the 'cloud burn'.

An indication of the devastation that may be caused by these overpressures is described in the following accident box.

ACCIDENT A major explosion occurred at a FCCU (fluid catalytic cracking unit). Six operators were killed, three of whom were in the control room. Two others were seriously injured. The FCCU and surrounding process units were severely damaged, resulting in the whole refinery being shut down for several months, with the FCCU taking a year to rebuild and return to service. The total cost of the incident was around $600 millions of which 64% was attributed to repair to the facilities, 34% loss of production, and 2% civil liability.

The incident arose from a release of about 15 tonnes of hydrocarbons when a hole of about 25 cm^2 (about 4 square inches) in area, created through corrosion, suddenly appeared in a 200 mm (8 inch) absorber stripper reflux cooler bypass line. The release occurred over about 10 minutes creating a vapour cloud of 14,000 m^2 (150,000 square feet), which engulfed other process units. When the cloud ignited on a furnace, the UVCE caused massive damage to plant and equipment and collapsed the roof on the process control room.

Figure 80 *Massive destruction caused by a vapour cloud explosion.*

A massive fire created a major domino effect that led to further loss of containment and escalation. In addition to the FCCU gas recovery plant, four main areas were affected including a main pipe rack, a building containing a turbine generator, a 2,500 m^3 (15,700 bbls) spent caustic soda tank containing a layer of light hydrocarbon, and a 5,000 m^3 (31,500 bbls) fuel oil tank. Although the fires in the turbine building and tankage were extinguished quickly, other areas were allowed to burn under control for three days until their sources of fuel were exhausted. The control room was built in 1953, and was not blast protected.

BLEVE

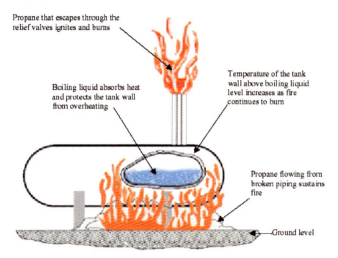

Figure 81 The BLEVE condition (Drawing U. S. Chemical Safety and Hazard investigation Board).

The BLEVE condition will develop where a pressurized LPG container or light ends container is subjected to flame impingement. The heat input raises the temperature of the liquid within until the liquid boils and vapour pressure increases until a relief valve opens. The relieved vapour usually ignites. Continuous flame impingement means continuous boiling liquid which will be burned off, thus causing dropping liquid level.

When the steel shell no longer has liquid within to cool the flame impinged area of steel, the temperature of the steel rises to failure point and the shell fails (tears) open, releasing the contents in a mass of combined burning vapour and boiling liquid that expands to produce more vapours to feed the resultant fireball.

Note: LNG tanks are unlikely to BLEVE because:

- tanks are double-walled and well-insulated or vacuum-jacketed;
- the outer shell will prevent direct flame impingement on the inner tank;
- the insulation between the outer and inner wall will greatly slow heat transfer to the LNG.

Thermal radiation levels from the resulting rising fireball are sufficiently high and wide to result in fatalities, serious burn injuries, and damage to plant and equipment.

In fighting fires where liquid butane and propane are stored in horizontal cylindrical pressure vessels, operators should be aware that these vessels sometimes fail at a circumferential seam. When this occurs, the sudden release of pressure sends the vessel off like a rocket (Figure 82). No one should be positioned in front of either end of such a vessel during a fire.

Figure 82 Light ends pressure vessels may become rockets if they fail at circumferential seams.

However, this does not mean that staying near the sides is safe either as during a BLEVE, tank debris can fly in any direction, not just from the ends.

ACCIDENT A propane tank fire started after two teenagers driving an all-terrain vehicle ploughed into unprotected propane piping at a farm. This above ground piping ran from the propane storage tank to vaporizers, which fueled heaters located in barns and other farm structures. The 42-foot long, cigar-shaped storage tank contained propane liquid and vapour under pressure, and the tank was about half full at the time of the incident. The collision severed one pipe and damaged another, triggering a significant propane leak under the 18,000 gallon tank. About five minutes later, propane vapour leaking from the damaged pipes ignited and burst into flames, engulfing the tank and beginning to heat the propane inside.

Because of the flames, arriving firefighters could not approach a manual shut-off valve to stop the propane leak, and they decided to let the tank fire burn itself out. The fire chief on the scene believed that in the event of an explosion, fragments would be thrown from the tank's two dome shaped welded ends. The areas near the sides of the tank, he believed, would be relatively safe. Shortly after their arrival, firefighters approached the sides of the flaming tank and began spraying the surrounding buildings to prevent the spread of fire.

Just seven minutes later, the burning propane tank ruptured completely, experiencing a Boiling Liquid Expanding Vapour Explosion or BLEVE.

Figure 83 A piece of the 18,000-gallon tank inside a turkey barn (picture, U. S. Chemical Safety and Hazard Investigation Board).

continued

The propane tank was blown into at least 36 pieces, some of which flew 100 feet or more. Some of the shrapnel struck firefighters, killing two of them. Other pieces smashed into buildings, leaving nearly $250,000 in property damage.

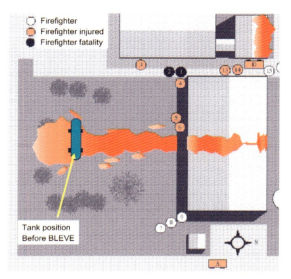

Figure 84 Firefighters' positions at the time of the BLEVE (Illustration NFPA).

Figure 85a Typical fireballs from BLEVE events.

Figure 85b LPG vessel failures under fire.

The only strategy for response to an impending BLEVE is to evacuate to a safe distance. A BLEVE will typically form a fireball although this will be a short-lived event. Like a flash fire, the duration of the fireball, by itself, is normally not sufficient to cause steelwork failures. However, any vapours present in the fireball area will ignite and may lead to secondary vent fires or other fire events.

Note that vessel failure under fire can occur with any size of LPG vessel. Figure 85b shows an example of a 33 kg LPG bottle which ruptured after being involved in a fire, a depot and a LPG truck during unloading.

Personal Protective Equipment (PPE) for responders
Selection of PPE

Responder PPE should always meet an EN, NFPA or equivalent standard. PPE should be viewed as a 'system' in terms of total wearer protection, rather than separate items that can be mixed to suit individual budgets.

The level of responder intervention and exposure will determine the levels of PPE required. For instance, at the lower exposure end, if the responder actions are simply to actuate a fire system switch and check evacuation status, then the PPE selection may be for only normal facility coveralls (which will probably be fire resistant in any case). However, where responder actions require manual deployment of fire monitors, foam monitors, use of breathing apparatus, etc, there will be a need for 'fire turnout gear'. This will typically consist of:

- fire helmet with visor;
- fire coat;
- fire trousers;
- fire boots;
- fire gloves.

The use of fire resistant (see below for clarification) materials for such PPE is critical to responder protection. In addition, responder normal workwear needs to consider fire resistant coveralls to ensure maximum protection is provided under fire exposed conditions. This becomes obviously important when responders may be working nearby to a gas release (knowingly or otherwise), where there is potential for a flash fire event.

Fire retardant treatments versus inherently fire resistant

Materials that are inherently fire resistant, such as Nomex fibres, offer continuous fire resistance for responders. This resistance can be provided in coveralls, two piece protection suit or fire turnout gear. There are materials such as cotton that have been treated with fire retardant chemicals to offer fire protection capability but such material, when subjected to heat or flame, will often result in the mass release of these chemicals thus making the garment unfit for further service.

Additionally, the regular cleaning of fire retardant treated clothing will result in washout of the protective chemicals. Therefore, the preference should be for inherently fire resistant clothing rather than treated clothing.

Scenario-specific emergency response plans

Although facilities should have emergency procedures in place for higher level control of an incident, generic and specific Emergency Response Plans (ERPs) should be prepared for credible serious or major incidents at facilities.

The requirement to prepare an emergency response plan at major hazard sites, such as many of those that will be handling light ends, is contained in regulations such as OSHA PSM in the USA or SEVESO in Europe (COMAH in UK).

The ERPs should be:

- based on facility potential credible serious or major scenarios;
- relevant to the facility systems and equipment (site specific);
- fit-for-purpose;
- easy to use;
- helpful to the end users.

Preferably, ERPs should consist of a single front page of text intended as guidance and instruction for incident responders whilst on the reverse of the text page, an 'effects' map is provided.

This effects map should indicate either the potential jet or pool fire flame impingement area, radiant heat hazard areas or other affected area where adjacent or nearby plant, tanks, vessels and associated equipment will or may be affected by an emergency incident.

Unignited gas releases contours should also be indicated on what may be termed 'effects' maps to give an appreciation of the potential worst case scale and severity of an incident.

ERP's purpose is to provide instant written instructions, guidance and helpful information for personnel to assist them at the critical early stage of a serious or major incident, and to provide sufficient potential hazard information to enable informed decisions on the safety of personnel responding to the incident.

The ERPs are intended to provide guidance for the first 20 to 30 minutes of the incident and indicate the actions and resources required to deal with the incident during this time. Once this period of time elapses, a stable response should have been established and if the incident duration should be prolonged, an ongoing strategy for dealing with this should be developed by those managing the incident.

Figure 86 on page 85 is an example of a two page ERP with text page and effects map, in this case, for an un-ignited gas release.

SAFE HANDLING OF LIGHT ENDS

ERP-1	PROPANE SPHERE	Date:
EMERGENCY RESPONSE PLAN FOR:	JET FIRE FROM SPHERES, VALVES, EQUIPMENT OR RPING	Approved: Rev: Date:

STRATEGY Confirmation of fire event – Affected area ESD – Non essential personnel evacuated – CCB alerts road car loading – CCB ensures CFS has received alarm – CFS arrival and earliest assessment of flame impingement – CFS/OSC mobilise for incident level – CFS decide if necessary and safe to cool muster affected steelwork/exposures or if bund piping and valves need immediate cooling or if full evacuation is required – CSFB arrive and receive ERP – CSFB assess incident conditions – If required necessary and safe to do so, water streams to be directed on to sphere vessels stairways, access towers and top valves etc – If sphere affected fire incident duration greater than 2 hours, additional water monitors to be deployed – Once all cooling operations are in place, all fire responders retire to a safe distance until fire burns out.

1st RESPONDERS	ACTIONS	EQUIPMENT/RESOURCES	INFO/COMMENTS
[ ] CCB Technician	Alert PLPG operator for confirmation.	Radio or AB Telephone	
[ ] CCB Technician	Initiate affected area ESD on confirmation.	PLPG ESD Controls	
[ ] CCB Technician	Alert road car loading to incident.	Telephone Emergency Number	Available plant technicians to act as Local Evacuation Officers [LEO's] and to direct personnel to a safe and appropriate muster point.
[ ] CCB Technician	Alert SAS/SOL to incident.	Radio or Telephone	
[ ] CCB Technician	Ensure fire station are responding.	Radio or CFS Fire Control Room	
[ ] CCB Technician	Ensure OSC and/or PLPG Technicians have carried out plant evacuation.	Radio or Telephone	
[ ] CCB Technician	Ensure all these section actions are checked and advise OSC of section completion.	Radio the BP OSC and advise all items complete.	
[ ] PLPG-OSC	Make contact with CFS officer and confirm all this section items are completed.		

2nd RESPONDENERS	ACTIONS	EQUIPMENT/RESOURCES	INFO/COMMENTS
[ ] CFS Fire Officer	Contact OSC and confirm incident level.	Radio through to CFS fire station control room.	Stairways & sphere tops have unprotected steel work and valves which can conduct heat and distout. Bund valves & piping have only 15 minutes jet fire protection so cool urgently if flame impinged.
[ ] CFS Fire Officer	Immediate check on flame impingement or radiant heat on piping valves in bunds.	Check fire hazard contours map on reverse page for potential radiant heat. Bund equipment may need immediate cooling.	
[ ] CFS Fire Officer	Check fire/PFP conditions on sphere and if safe & necessary to cool exposures.	Check fire hazard contours map on reverse page.	
[ ] CFS Fire Officer	If cooling needed for a sphere, direct water to stairways and top access platforms.	Minimum 2 × foam tenders, 5 CFS responders, 3 × 4500 lpm water monitors, 24 × 70mm delivery hose.	
[ ] CFS Fire Officer	Check all actions in this section are marked and give ERP to CSFB Officer on arrival.	This ERP.	Inform CSFB Fire Officer that all listed actions have been carried out up to this point.

3rd RESPONDERS	ACTIONS	EQUIPMENT/RESOURCES	INFO/COMMENTS
[ ] CSFB Fire Officer	Contact OSC and CFS F.O. on arrival and assess if safe to continue cooling operations or if more cooling required.	Minimum 2 × water tenders, 8 CSFB responders, 3 × 4500 lpm water monitors, 24 × 70mm delivery hose if required.	Water monitors are available from either the general PLPG area or from the CFS fire tenders.
[ ] CSFB Fire Officer	Review escalation potential and request plans for escalation or evacuation	Refer RTF ERP-1 for Blend Tank T-293 potential effects.	
[ ] CSFB Fire Officer	Maintain BP OSC liaison for updates. Monitor PFP conditions	Refer PLPG ERP-10 & 11 for Butane Drier potential effects.	OSC can advise time of fire, sphere inventory levels, possibility of product transfer, possible duration of incident etc. Safe distance must consider BLEVE/Fireball impact shown in ERP-12.
[ ] CSFB Fire Officer	Once all cooling operations in place retire to a safe distance and await burn out.	Refer PLPG ERP-12 for BLEVE/Fireball potential effects. (See Other Concerns below)	

INCIDENT POTENTIAL HAZARDS The sphere bundpiping and ESD valves in sphere bunds have only 15 minutes PFP for jet fire conditions so urgent cooling needs to be applied. Flame or radiant heat affected sphere stairways, stair access tops and valves need cooling since they can conduct heat to sphere steelwork below PFP. Stairways can distort, causing metal stress, fatigue and failure. Spheres have PFP rated as 2 hours jet/pool fire protection but if fire event will be greater than 2 hours, additional water monitors should be organised in readiness, if safe to do so, in case spheres need cooling. If deterioration of PFP is noted, additional cooling water may be required to protect exposed steelwork underneath. If water supplies inadequate it may be necessary to fully evacuate area that may be affected by a BLEVE. If relatively low pressure jet fire, do not try to extinguish unless there is a clear incident control benefit without greater risk.

OTHER CONCERNS Pressure water jets applied on to PFP may damage the coating so exercise care to ensure cooling water used as spray application onto PFP unless the PFP has or is degrading. Any doubts on fire affected sphere integrity should consider BLEVE/Fireball **Refer to PLPG ERP-12 for BLEVE/Fireball potential effects.**	RADIOACTIVE HAZARDS / ASBESTOS HAZARDS / TOXIC HAZARDS Asbestos confirmed presence in sphere support legs. Consider all sphere support legs as having an asbestos hazard. No radioactive sources at the PLPG. No Toxic Hazards.

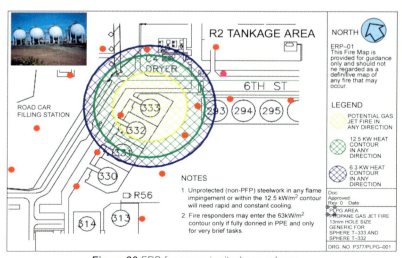

Figure 86 ERP for an un-ignited gas release.

SAFE HANDLING OF LIGHT ENDS

10

Some points to remember

1. Light ends are volatile. If the pressure is released from liquid light ends, they vaporize rapidly.

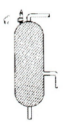

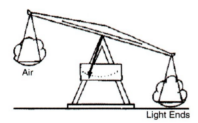

2. Most light ends are heavier than air.

3. Hydrocarbon vapour-air mixtures must be controlled. There will be a source of ignition.

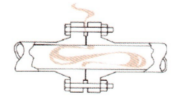

4. The viscosity of liquid light ends is low. This makes them difficult to contain.

5. Heavier hydrocarbons act like light ends when they are hot.
6. Sampling of liquid light ends can be hazardous. Be sure proper equipment and procedures are used.

SAFE HANDLING OF LIGHT ENDS

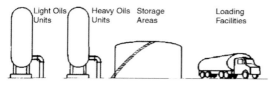

7. Light ends are everywhere in refineries.

8. During shutdowns, remove residual hydrocarbons before admitting any air. During start-ups, remove residual air before admitting hydrocarbons. Keep air out when on-stream.

9. Check for oxygen in vapour spaces of nonvented operating drums and storage vessels. Make certain that tank cars and truck transports do not contain oxygen before loading them.

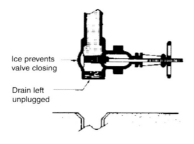

10. Drains may freeze when depressuring equipment or drawing water. Be sure that ice is not stopping the flow.

11. Watch for leaks. A small liquid leak becomes a large dangerous vapour cloud.

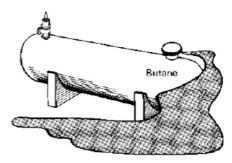

1 cubic foot (7.5 gallons) of liquid = 225 cubic feet of gas

87

SAFE HANDLING OF LIGHT ENDS

12. Do not allow uncontrolled hot work where light ends are processed.

13. Keep flare stack knockout drums drained.

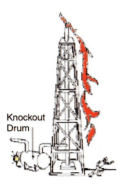

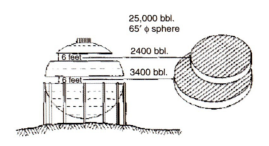

14. Remember that each foot on the level gauge often does not represent the same volume.

15. Do not overfill. Liquid expansion may cause relief valves to pop.

16. Do not pressure a container with air to empty it. Use nitrogen.

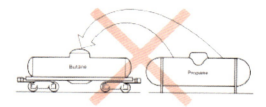

17. Be sure the proper storage or transport container is used for the hydrocarbon to be contained.

18. Inspect and hydrostatically test LPG hoses periodically.

19. Never leave a connected tank car or truck transport unattended.

20. If an LPG spill or leak occurs, clear the area of any source of ignition. Stop the leakage or flow of LPG to the affected equipment.

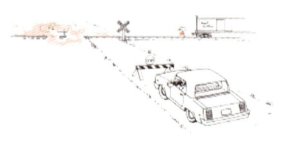

21. Vessels containing LPG that are exposed to fire should be sprayed with water to keep the shell surfaces cool.

Bibliography

- API
  - Std. 2510, Design and Construction of Liquefied Petroleum Gas (LPG) Installations
  - RP 2510A Fire Protection Considerations for the Design and Operation of Liquefied Petroleum Gas (LPG) Storage Facilities.
- LP Gas Association: codes of practice
  1. Bulk LPG Storage at Fixed Installations
  2. Safe Handling and Transport of LPG in Road Tankers and Tank Containers by Road
  3. Prevention or Control of Fire Involving LPG
  12. Recommendations for Safe Practice in the Design and Operation of LPG Cylinder Filling Plants
  14. Hoses for the Transfer of LPG in Bulk Installation, Inspection, Testing and Maintenance
  15. Valves and Fittings for LPG Service
  17. Purging LPG Vessels and Systems.
- ASME B31.3 Process Piping Code
- UK Health and Safety Executive (HSE):
  - The Storage of LPG at fixed installations, HS(G)34
  - The bulk transfer of dangerous liquids and gases between ship and shore, HSG 186
  - The loading and unloading of bulk flammable liquids and gases at harbours and inland waterways, GS40.
- UK Energy Institute: Model Code of Safe Practice:
  Part 19 – Fire Precautions at Petroleum Refineries and Bulk Storage Installations
  Part 9 – Bulk Pressure Storage and Refrigerated LPG.
- National Fire Protection Association:
  - NFPA 15: Water Spray Fixed Systems for Fire Protection
  - NFPA 58: Standard for the Storage and Handling of Liquefied Petroleum Gases.
- SIGTTO:
  - Liquefied Gas Handling Principles on Ships and in Terminals
  - Liquefied Gas Fire Hazard Management

	- A Guide to Contingency Planning for Marine Terminals Handling Liquefied Bulk Gases
	- Cargo Firefighting on Liquefied Gas Carriers
	- Safety in Liquefied Gas Marine Transportation and Terminal Operations (CD)
	- Liquefied Gas Carriers: Your Personal Safety Guide.
- ICS: Tanker Safety Guide (Liquefied Gas)
- OCIMF: Barge Safety (Liquefied Cargoes in Bulk)
- IMO: International Code for the Construction and Equipment of Ships Carrying Liquefied Gases in Bulk (IGC Code)

Appendix 1: Example of operator fire safety checklist for LPG storage

The following is an example of a checklist for LPG spheres and vessels to be used by operators for their LPG storage facilities. It also includes maintenance aspects as well as fire-related inspection checks and is not exhaustive.

A1.1 Technician safety checks

The following inspections and checks should be carried out by technicians or other operations personnel with responsibility for operating LPG facilities. The frequency is based on good international industry practice. Technicians or operators are fully expected to report any conditions or defects that do not conform to the purpose or intent of the inspection or checks.

Daily

Bund drains: Functional open/close test and ensure that valves are left in the fully closed position.

Dewatering/sample locked valves: Check required valves are locked or otherwise held in the open position.

Weekly

Pressure relief valves: Check block valves on PRVs are locked or otherwise held in the open position.

Monthly

Pressure relief valves:

- visual inspection of vent covers for damage/wear/corrosion;
- visual inspection of vent tail pipe drain for corrosion and to ensure it is not blocked or clogged;
- visual inspection for corrosion/wear.

Sphere/vessel valves:

- functional open/close test using remote control panel switches;
- visual inspection to ensure closed or open valves are correctly indicated at remote operating panels;
- visual inspection of remote status panel lights.

Small bore screw fittings:
- check for corrosion or signs of thread wear and tear;
- check for minor leaks.

Sampling points:
- check position of valves;
- check for minor leaks.

Flanges:
- check for corrosion or signs of wear and tear;
- check for minor leaks.

All liquid and vapour pipelines and drain points:

Visual inspection to ensure piping and drain points are not submerged due to rain fall in the bund floor or due to ground erosion of piping supports causing damp/corrosion conditions directly at piping or drain valves.

Pumps:
- visual checks for general good condition and vibrations;
- checks for indications of minor leaks at pump seal area, ancillary systems such as seals, lubeoil, etc.;
- visual inspection of supports for small bore piping associated with pumps for damage, corrosion or general wear and tear.

Support legs:

Visual inspection:
- for fireproofing delamination, cracks, flaking or other damage;
- for signs of corrosion from steelwork under the fireproofing;
- of cross tensioners for corrosion or wear and tear.

Access lighting:
- visual inspection of light cover seals for signs of rain or water entry indicating seal degradation and potential gas/vapour ingress.

Electrical equipment and fittings:
- visual inspection of covers, seals and wiring for corrosion, integrity, wear and tear.

Earthing/bonding:
- Visual inspection for corrosion, wear and tear at the cable and fixing points on bullet or sphere or sphere legs and ground.

Grass, brush, vegetation, trees:
- check grass or brush or vegetation in or around bunds is cut short to prevent fire risk;
- check tree roots are not damaging bund walls.

Signage:

- check presence and legibility of valve and pipe identification;
- check presence and legibility of warning signs, safety signs and fire systems signs.

A1.2 Fire protection and fire safety checks

The following checks should be either made or arranged by the fire response, fire safety team, or fire safety representative, depending on the location of the LPG facility.

Quarterly

Gas detection:

- test for LFL actuation and alarm functions and any executive actions;
- re-calibration of detectors.

6 monthly

Water deluge system:

- visual inspection of piping and manual valves for damage, corrosion and general good condition;
- functional testing of water deluge to ensure operation and total sphere/bullet cooling water total coverage;
- visual inspection for partially or wholly blocked spray nozzles;
- functional testing of water deluge system drains.

Water filling system:

- visual inspection for piping condition, corrosion and damage or wear and tear;
- valve operation and flushing;
- visual inspection of connecting fire hose.

Firewater hydrants and monitors:

- functional test for flow and valve operation;
- functional test for flow, stream pattern adjustments and valve operation;
- visual inspection for corrosion, damage or general wear and tear.

Annually

Water filling hose:

- pressure testing of connecting fire hose.

2 Yearly

Gas detection:

- change sensors (Catalytic sensor types).

Test yourself!

1. Methane, ethane, propane and butane are gases at atmospheric temperature and pressure.

   True ☐ False ☐

2. Catalytic gas indicators can perform well in inert atmospheres.

   True ☐ False ☐

3. Natural and fuel gases which are mostly methane are lighter than air at ambient temperature.

   True ☐ False ☐

4. Ethane, propane and butane are lighter than air at ambient temperature.

   True ☐ False ☐

5. Vessels can safely be filled to 100% with liquid LPG.

   True ☐ False ☐

6. Draining water from a LPG vessel is an operation which can be left unattended.

   True ☐ False ☐

7. Pumps with double mechanical seals have the highest degree of protection against leaks.

   True ☐ False ☐

8. A closed circulating pipe system should be used to blend butane in gasoline.

   True ☐ False ☐

9. During a fire, it is safe to approach a cigar-shaped vessel from the sides.

   True ☐ False ☐

10. A jet fire has the potential to cause rapid escalation.

    True ☐ False ☐

ANSWERS
1T/2F/3T/4F/5F/6F/7T/8T/9F/10T

Acronyms and abbreviations

CO_2	Carbon Dioxide
ERP	Emergency Response Plan
FCCU	Fluid Catalytic Cracker Unit
LNG	Liquefied Natural Gas (mainly methane)
LPG	Liquefied Petroleum Gas (Propane–Butane)
O_2	Oxygen
PPE	Personal Protective Equipment
PRV	Pressure Relief Valve
WSE	Written Scheme of Examination

Your notes